Maize Cultivation and Improvement

Maize Cultivation and Improvement

Guest Editor

Glauco Vieira Miranda

MDPI

Basel • Beijing • Wuhan • Barcelona • Belgrade • Novi Sad • Cluj • Manchester

Guest Editor
Glauco Vieira Miranda
Department of Agronomy
Federal Technological
University of Paraná
Santa Helena
Brazil

Editorial Office
MDPI AG
Grosspeteranlage 5
4052 Basel, Switzerland

This is a reprint of the Special Issue, published open access by the journal *Plants* (ISSN 2223-7747), freely accessible at: https://www.mdpi.com/journal/plants/special_issues/35AE0225C8.

For citation purposes, cite each article independently as indicated on the article page online and as indicated below:

Lastname, A.A.; Lastname, B.B. Article Title. *Journal Name* **Year**, *Volume Number*, Page Range.

ISBN 978-3-7258-7360-9 (Hbk)
ISBN 978-3-7258-7361-6 (PDF)
https://doi.org/10.3390/books978-3-7258-7361-6

Contents

About the Editor

Glauco Vieira Miranda

Glauco Vieira Miranda is a Professor in the Department of Agronomy at the Federal University of Technology Paraná in Brazil. His research focuses on the integration of plant breeding, genomics, artificial intelligence, and data-driven agriculture to improve crop productivity and resilience. He works on the development of intelligent decision support systems, high-throughput phenotyping strategies, and predictive models that combine genomic, phenomic, and environmental data to accelerate genetic gain in major crops. He leads interdisciplinary research initiatives aimed at advancing sustainable agricultural systems, including projects related to genomic prediction, crop stress tolerance, and digital agriculture platforms. Miranda has coordinated large collaborative research networks involving universities, research institutes, and private-sector partners, contributing to the development of innovative technologies and analytical frameworks for modern crop improvement. His work has resulted in numerous scientific publications, patents, and technological innovations in the fields of plant breeding and computational agriculture. He has also played an active role in graduate education and international scientific collaboration, promoting the integration of biotechnology, artificial intelligence, and agronomic sciences to address global challenges in food production and agricultural sustainability.

plants

Editorial

Maize Cultivation and Improvement

Glauco Vieira Miranda

Department of Agronomy, Federal University of Technology Paraná, Santa Helena 85892-114, Brazil; glaucovmiranda@utfpr.edu.br

This Special Issue "Maize Cultivation and Improvement" gathers eight high-quality contributions that collectively reflect the contemporary landscape of maize research. These articles span molecular genetics, quantitative breeding, stress physiology, proteomics, genomic prediction, agronomic management, artificial intelligence, and meta-analytical synthesis. Together, these articles illustrate how modern maize improvement is increasingly interdisciplinary, integrating biotechnology, data science, and field-based validation to accelerate genetic gain and enhance sustainability.

Maize (*Zea mays* L.) remains one of the most strategically important crops worldwide, serving as food, feed, fiber, and fuel. However, climate variability, increasing biotic and abiotic stresses, and the demand for higher productivity under resource constraints require transformative approaches. The studies compiled in this volume demonstrate how germplasm development, precision phenotyping, and advanced analytics can converge to address these global challenges.

1. Molecular Regulation and Functional Genetics

Understanding the genetic and molecular mechanisms underlying key agronomic traits remains foundational to crop improvement. Two studies in this Special Issue provide important advances in this direction.

The identification of a single-base mutation in the *Dwarf 1* (D1) gene responsible for reduced plant height offers valuable insight into plant architecture control [1]. Plant height directly influences lodging resistance, biomass allocation, and yield stability. The identification of this mutation provides a potential target for marker-assisted selection and functional validation strategies aimed at optimizing canopy structure in diverse environments.

Similarly, the investigation of cysteine protease ZmPCP in regulating pollen viability, pollen tube growth, and seed development addresses reproductive resilience under drought and heat stress [2]. Reproductive success is one of the most climate-sensitive phases in maize production. By linking molecular regulation to pollination efficiency and seed formation, the study contributes to the development of germplasm that is better adapted to extreme environmental conditions.

2. Stress Physiology and Proteomic Responses

Climate change intensifies the frequency of combined drought and heat stress events. The comparative leaf proteome analysis of maize under simultaneous drought and heat stress provides a system-level understanding of stress tolerance mechanisms [3]. The screening of 45 inbred lines and identifying contrasting proteomic responses highlights the importance of integrating physiological evaluation with omics technologies. Such approaches enable the identification of candidate pathways and biomarkers for stress resilience.

The work reinforces the necessity of coupling molecular-level insights with breeding strategies, allowing selection decisions to move beyond purely phenotypic evaluation toward mechanistic understanding.

3. Quantitative Breeding and Genetic Gain

Sustained genetic progress depends on efficient selection methodologies. A study on genetic gains with a field validation of synthetic tropical maize populations demonstrates the power of selection indexes combined with REML/BLUP methodology [4]. By evaluating recurrent selection strategies under tropical conditions, the study underscores the importance of robust statistical modeling in managing genotype–phenotype interactions and inbreeding effects.

Complementing this, a work on enhancing across-population genomic prediction for maize hybrids addressed a central limitation of genomic selection: the prediction accuracy across heterotic groups and breeding populations [5]. Improving cross-population prediction expands the operational scope of genomic selection and reduces breeding cycle length, directly contributing to accelerated genetic gain.

Together, these studies illustrate the evolution from classical recurrent selection to data-driven genomic breeding pipelines, reinforcing the integration of statistical genetics and computational methodologies in modern maize improvement programs.

4. Agronomic Management and Environmental Interactions

Genetic potential can only be realized through optimized management. A study evaluating the effects of semi-arid environmental conditions and agronomic traits on grain quality emphasizes the interaction between genotype and environment in determining both yield and grain composition [6]. As production expands into marginal or stress-prone regions, understanding how environmental variables influence quality parameters will become increasingly important for food and feed industries.

At a broader scale, the meta-analysis on increased planting density in Northeast China synthesizes evidence from 508 paired observations [7]. By quantitatively assessing yield responses to planting density adjustments, the work provides a statistically robust framework for optimizing management strategies. The integration of large-scale evidence through a meta-analysis strengthens agronomic recommendations and supports region-specific decision-making.

5. Digital Agriculture and High-Throughput Phenotyping

One of the most forward-looking contributions in this Special Issue is a CEHD framework for detection and height estimation of fresh corn ears under field conditions [8]. By applying deep learning architectures (YOLO-based models) for real-time ear detection and height estimation, the research exemplifies the integration of artificial intelligence into crop production systems.

Precision detection supports dynamic harvester adjustment, reduces mechanical damage, and improves harvest efficiency. More broadly, the study signals the transition toward Industry 4.0 in maize production, where visual computing, machine learning, and real-time analytics enhance operational precision and reduce post-harvest losses.

6. Convergence of Technologies in Maize Improvement

Across all eight papers, a clear theme emerges: the convergence of biotechnology, quantitative genetics, omics platforms, high-throughput phenotyping, and artificial intelligence.

The key integrative advances highlighted in this volume include:

- Functional gene identification linked to agronomic performance.
- The proteomic and molecular characterization of stress tolerance.
- REML/BLUP and selection indices to enhance recurrent selection efficiency.
- Genomic prediction models extending across breeding populations.
- Evidence-based agronomic optimization via meta-analysis.

 https://doi.org/10.3390/plants15050794

- AI-driven field phenotyping for harvest precision.

This convergence allows shortened breeding cycles, reduced experimental population size requirements, enhanced prediction accuracy, and strengthened translation of genetic knowledge into field performance.

7. Future Directions

Looking ahead, maize improvement will increasingly rely on the following:

1. Multi-omics integration (genomics, transcriptomics, proteomics, phenomics).
2. Climate-resilient germplasm development targeting combined stress scenarios.
3. Scalable genomic prediction models across heterotic groups.
4. Digital agriculture platforms enabling real-time crop monitoring.
5. Data-driven agronomic optimization supported by large-scale synthesis approaches.

The integration of these domains will be critical for achieving sustainable yield increases while maintaining environmental stewardship and economic viability.

8. Concluding Remarks

The contributions compiled in this Special Issue reflect the dynamism in maize research and the collaborative efforts of scientists worldwide. By combining molecular biology, advanced statistics, computational intelligence, and field validation, these studies collectively advance the theoretical and applied dimensions of maize cultivation and improvement.

I would like to sincerely thank all authors, reviewers, and the editorial team for their dedication and scientific rigor. The success of this Special Issue demonstrates the vitality of interdisciplinary research in maize and sets the foundation for continued innovation in germplasm development, breeding strategies, and sustainable production systems.

The publication of this Special Issue in book format provides a consolidated reference for researchers, breeders, agronomists, and students engaged in advancing maize science. I hope that the insights presented here will contribute meaningfully to global food security and to the continued evolution of maize breeding in an era defined by climatic uncertainty and technological transformation.

Conflicts of Interest: The author declares no conflict of interest.

References

1. Wang, P.; Liang, B.; Li, Z.; Dong, H.; Zhang, L.; Lu, X. The Identification of a Single-Base Mutation in the Maize *Dwarf1* Gene Responsible for Reduced Plant Height in the Mutant 16N125. *Plants* **2025**, *14*, 1217. [CrossRef] [PubMed]
2. Li, Y.; Wang, W.; Liu, H.; Wang, W. Regulation of Pollen Viability, Pollen Tube Growth and Seed Development in Maize by Application of Cysteine Protease ZmPCP. *Plants* **2026**, *15*, 677. [CrossRef]
3. Pfunde, C.; Mutengwa, C.S.; Bradley, G.; Chiuta, N.E. Comparative Leaf Proteome Analysis of Maize (*Zea mays* L.) Exposed to Combined Drought and Heat Stress. *Plants* **2025**, *14*, 3419. [CrossRef] [PubMed]
4. Sousa, A.M.d.C.B.d.; Resende, M.P.M.; Crispim-Filho, A.J.; Miranda, G.V.; Reis, E.F.d. Genetic Gains and Field Validation of Synthetic Populations in Tropical Maize Using Selection Indexes and REML/BLUP. *Plants* **2025**, *14*, 3149. [CrossRef] [PubMed]
5. Yu, G.; Li, F.; Wang, X.; Zhang, Y.; Zhou, K.; Yang, W.; Guan, X.; Zhang, X.; Xu, C.; Xu, Y. Enhancing Across-Population Genomic Prediction for Maize Hybrids. *Plants* **2024**, *13*, 3105. [CrossRef] [PubMed]
6. Bongianino, N.F.; Steffolani, M.E.; Morales, C.D.; Biasutti, C.A.; León, A.E. Semi-Arid Environmental Conditions and Agronomic Traits Impact on the Grain Quality of Diverse Maize Genotypes. *Plants* **2024**, *13*, 2482. [CrossRef] [PubMed]

 https://doi.org/10.3390/plants15050794

7. Zhang, J.; Wang, X.; Li, Y.; Yu, Z.; Zhang, R.; Yin, B.; Wang, H. A Meta-Analysis of the Effects of Increased Planting Density on Maize Yield in Northeast China. *Plants* **2026**, *15*, 544. [CrossRef] [PubMed]
8. Wang, H.; Li, Y.; Fu, J.; Fu, Q.; Qiao, Y. CEHD: A Unified Framework for Detection and Height Estimation of Fresh Corn Ears in Field Conditions. *Plants* **2026**, *15*, 38. [CrossRef] [PubMed]

 https://doi.org/10.3390/plants15050794

plants

MDPI

Article

Regulation of Pollen Viability, Pollen Tube Growth and Seed Development in Maize by Application of Cysteine Protease ZmPCP

Yanhua Li [1,2], Wenkang Wang [2], Hui Liu [2,*] and Wei Wang [2,*]

1 Department of Criminal Science and Technology, Henan Police College, Zhengzhou 450046, China; liyanhua@hnp.edu.cn
2 College of Life Sciences, State Key Laboratory of High-Efficiency Production of Wheat-Maize Double Cropping, Henan Agricultural University, Zhengzhou 450046, China
* Correspondence: liuhuisw@henau.edu.cn (H.L.); wangwei@henau.edu.cn (W.W.)

Abstract

In the process of maize production, extreme meteorological conditions such as drought and high temperature are often the main environmental stress factors affecting pollination efficiency. Previous studies have shown that, under adversity, the germination rate of pollen grains on the filaments of female spikes directly affects the success rate of reproduction and ultimately determines the grain yield. This study focuses on a cysteine protease named ZmPCP. The expression of this protease in maize pollen is significantly higher than in other tissues, and its specific function has not been clearly defined. Its localization in the cell membrane or apoplast was further confirmed by transient transfection experiments and plasmolysis. The interaction between ZmPCP and ZmSNAP33 was verified by yeast two-hybrid technology and a GST pull-down experiment, indicating that ZmPCP may affect pollen germination and stress resistance by regulating vesicle transport. Secondly, by analyzing the pollen germination rate of maize inbred lines B104, *ZmPCP*-KO and *ZmPCP*-OE transgenic maize plants, we found that *ZmPCP* overexpression could significantly enhance pollen viability and pollen tube growth under drought stress. After 1 h of short-term drying treatment, the pollen germination rate of the *ZmPCP*-OE line was maintained at 44%, which was significantly higher than that of the other lines. In addition, the observation of pollen tube growth showed that *ZmPCP* overexpression could promote the extension of pollen tubes in the filament. Moreover, a transcriptome sequencing analysis revealed the regulatory effects of *ZmPCP* on pollen in multiple biological processes, including stress response, carbohydrate metabolism, growth and development, cell wall material metabolism, signal transduction, etc. The involved pathways of these differential genes indicate that *ZmPCP* enhances pollen drought tolerance and promotes pollen tube growth through a "metabolism signal structure". In the germination experiment on the seventh day, the germination rate of *ZmPCP*-OE maize seeds was the lowest, indicating that its overexpression inhibited seed germination. At the same time, *ZmPCP* overexpressing Arabidopsis showed a significant advantage in taproot growth under high-concentration ABA stress. ZmPCP provides an important theoretical basis for regulating the pollination process and improving the pollination efficiency of maize varieties through interaction with ZmSNAP33.

Keywords: pollen; viability; pollen tube growth; subcellular localization; yeast two-hybrid; seed development; maize

1. Introduction

Maize (*Zea mays* L.), as one of the world's three major staple crops, occupies a central position in China's agricultural economy. In 2024, its production reached approximately 295 million tons, with a year-on-year increase of 2.1% [1]. It serves not only as a source of human food and industrial raw materials but also as a critical source of livestock and poultry feed, playing an irreplaceable role in ensuring food security and promoting agricultural modernization [2]. However, influenced by global warming, the frequency of extreme weather events such as high temperatures and droughts has increased, significantly reducing maize cross-pollination efficiency. During the critical period of 10 to 15 days before flowering, the viability of pollen and silks is essential for kernel number, and this stage is highly vulnerable to abiotic stress interference [3]. Drought stress suppresses pollen viability because the loss of pollen moisture results in increased pollen mortality, associated with poor sucrose synthase activity [3]. In maize, two sucrose synthase (SUS) genes are downregulated in pollen under drought conditions, limiting the conversion of sucrose into D-fructose and thereby compromising pollen performance [4]. Successful fertilization in maize further relies on the rapid growth of pollen tubes, which grow at rates of up to 1 cm per hour through elongated styles (silks) before releasing their sperm cells within ovules [5]. Pollen tubes secrete numerous small proteins to facilitate communication with diverse maternal tissues during their transit from the stigma through the style's transmitting tract toward ovules. For example, rapid alkalinization factors (RALFs) regulate hydration as well as cell wall integrity during pollen germination, tube growth and reception in the model plant Arabidopsis [6]; *ZmRALF2/3* mutations disrupt pectin distribution, altering cell wall organization and thickness, which triggers pollen tube rupture [5].

Cysteine proteases (CPs) represent an important class of enzymes that are widely involved in pollen development [7], seed germination [8] and resistance to abiotic stresses [9]. Pollen coat proteins, including several CPs, are key components regulating pollen hydration, germination, and pollen tube elongation, with their molecular mechanisms being a research focus in plant reproductive biology [10]. In maize, for example, β-extensin EXPB1 binds to glucuronyl arabinoxylans, and its downregulation reduces pollen competitiveness [11]. In addition, the maize pollen coat also accumulates β-1,3-glucanase, endoxylanase, and exopolygalacturonase, which specifically hydrolyze cell wall polysaccharides [12,13]. ZmPCP, a pollen-specific CP cysteine protease belonging to the papain-like cysteine protease (C1-papain subfamily), is highly expressed in mature pollen. Extensive studies in multiple plant species have demonstrated crucial roles for CPs in pollen development. In Arabidopsis, *AtCEP1* is an important executor in the process of tapetal programmed cell death (PCD), and its overexpression leads to early tapetal PCD and pollen abortion [7]. Reduced expression of *AtCP51* results in early tapetum degeneration and defective pollen exine formation [14]. In tobacco, knockout of *NtCP56* causes delayed tapetal breakdown and the production of abortive pollen [15]. In rice, *OsCP1*, which is strongly expressed in anthers, is mutated by T-DNA insertion, leading to pollen degeneration [16]. Previous studies indicate that *ZmPCP* overexpression leads to abnormal morphology of pollen germination pores, reduced germination rates, and decreased fertility while simultaneously enhancing plant drought resistance [17]. Conversely, knockout experiments demonstrate the opposite effect, suggesting that *ZmPCP* plays a dual role in pollen development and stress response. In addition, the *ZmPCP*-OE and *ZmPCP*-KO maize lines used in this study were generated and characterized in our previous work [17], providing validated genetic materials for the functional analyses presented here.

Beyond reproductive tissues, CPs also play essential roles in seed development. In cereals, CPs mediate the degradation of seed storage proteins to release amino acids and nutrients that are necessary for seedling growth [8]. For instance, the *OsCP6* gene positively

https://doi.org/10.3390/plants15050677

regulates rice seed germination. The metabolism of endosperm storage and hydrolysis of storage proteins in the *oscp6-1* mutant decreased, and cysteine protease activity decreased under low-nitrogen conditions during early seedling development [18].

Moreover, CPs are important components of plant responses to abiotic stress, particularly drought and waterlogging [9]. In sweet potato, the overexpression of *SPCP3* in Arabidopsis increases drought sensitivity, whereas ectopic expression of *SPCP2* enhances drought tolerance [19].

Integrating these observations, we hypothesize that ZmPCP (*Zea mays* pollen-specific cysteine protease) modulates maize yield stability by differentially regulating reproductive development and stress-response pathways. The objectives of this study are to: (1) resolve molecular evolutionary relationships and sequence conservation of ZmPCP in *Zea mays*, *Zea mays* ssp. *mexicana (Mexican teosinte)*, and *Arabidopsis thaliana*; (2) characterize the ZmPCP protein interaction network through subcellular localization, yeast two-hybrid (Y2H) assays, and GST pull-down experiments; (3) evaluate transgenic plants under 28 °C drought stress to assess pollen viability, pollen tube elongation, and transcriptomic profiles (RNA-seq), thereby clarifying its regulatory role in pollen tube growth; and (4) investigate *ZmPCP* functions in seed germination and root development using transgenic maize and Arabidopsis models. This work aims to elucidate the core mechanism of ZmPCP in pollen germination, providing novel genetic targets for maize improvement to counteract climate-impaired pollination efficiency.

2. Results

2.1. Phylogenetic and Structural Analysis Reveals Evolutionary Proximity of ZmPCP Between Maize and Mexican Teosinte

To elucidate the phylogenetic relationships of *ZmPCP*, this study constructed a phylogenetic tree of cysteine proteases (CPs) among maize (*Zea mays*), its wild progenitor Mexican teosinte (*Zea mays* ssp. *mexicana*), and the model eudicot *Arabidopsis thaliana*. Based on functional and structural characteristics, the CP family was classified into nine subfamilies (Figure 1A). A phylogenetic analysis revealed that ZmPCP possesses conserved domains of the CP family, belonging to the THI group within the papain-like cysteine protease subfamily C1A. Notably, maize and Mexican teosinte CPs exhibit closer evolutionary proximity than those of *Arabidopsis thaliana* (Figure 1A). Further multiple sequence alignment of homologous proteins in the THI subgroup branch demonstrated 100% sequence identity between ZmPCP and Zmex03t011636_P01 from Mexican teosinte, while identity with Zmex01t001605_P01 reached 99.14% (Figure 1B). A conservation analysis confirmed that all these proteins contain the characteristic catalytic triad (Cys–His–Asn) of the C1A subfamily.

Protein tertiary structures predicted by AlphaFold2 indicated highly conserved spatial configurations among ZmPCP, Zmex03t011636_P01, and Zmex01t001605_P01 (Figure 2). The core domains (Inhibitor_29 and Peptidase _C1) of these proteins form a highly conserved core catalytic fold composed of tightly packed β-sheets and surrounding α-helices (colored in varying shades of blue). In contrast, the structural model of AtTHI1 exhibited noticeable deviations from ZmPCP, including differences in the overall fold architecture (Figure 2). This structural divergence corresponds with the low sequence identity between AtTHI1 and ZmPCP (40.17%) (Figure 1B), likely attributable to their distant phylogenetic relationship.

https://doi.org/10.3390/plants15050677

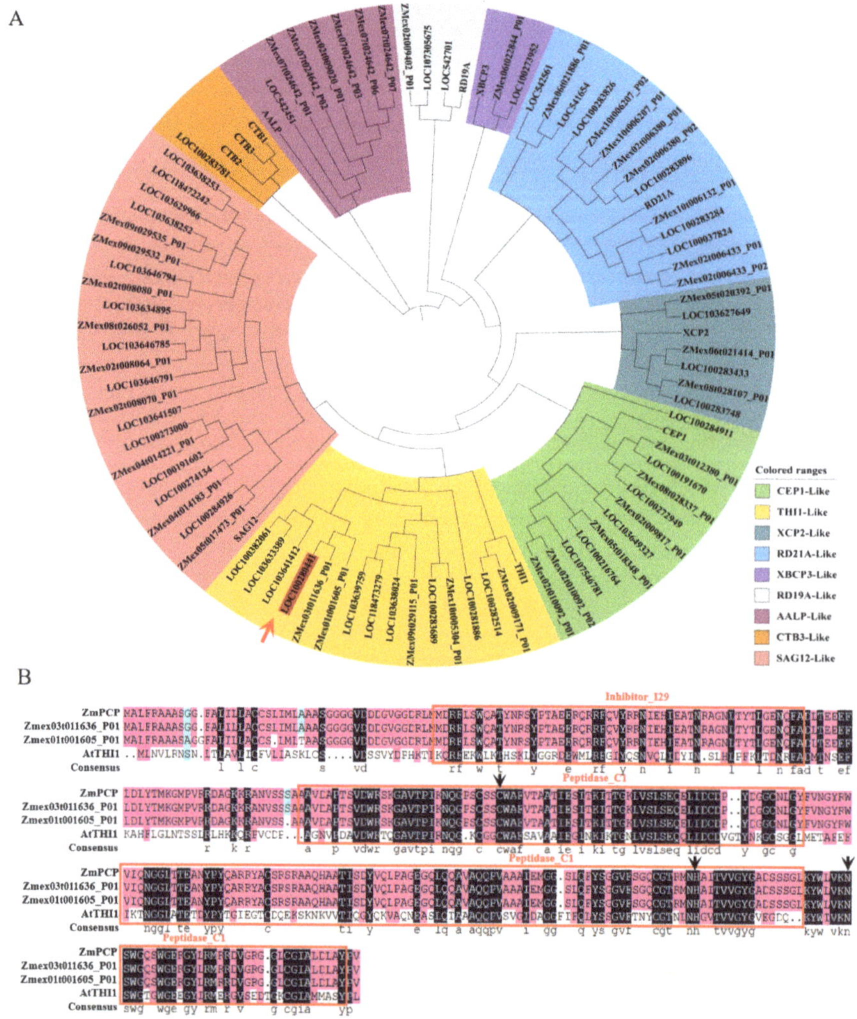

Figure 1. ZmPCP bioinformatics analysis. (**A**) Phylogenetic tree of ZmPCP in the cysteine protease family of maize, Mexican teosinte and *Arabidopsis thaliana*. The red arrow indicates ZmPCP. (**B**) Sequence alignment of ZmPCP with similar THI1 proteins. The black arrow indicates Cys, His and Asn.

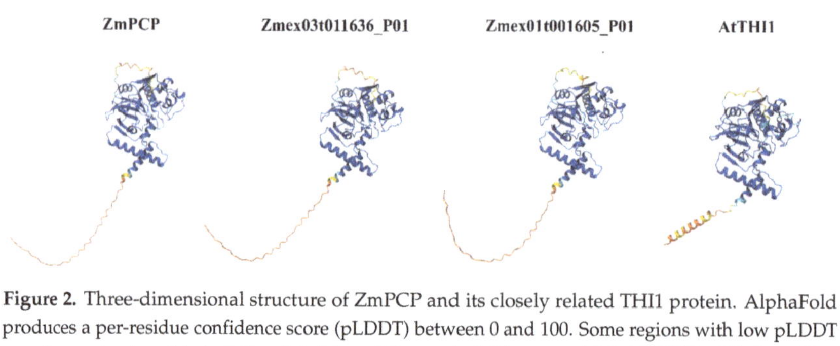

Figure 2. Three-dimensional structure of ZmPCP and its closely related THI1 protein. AlphaFold produces a per-residue confidence score (pLDDT) between 0 and 100. Some regions with low pLDDT may be unstructured in isolation. Model confidence: dark blue: very high (pLDDT > 90); light blue: confident (90 > pLDDT > 70); yellow: low (70 > pLDDT > 50); orange: very low (pLDDT < 50).

https://doi.org/10.3390/plants15050677

2.2. Subcellular Localization of ZmPCP

To determine the subcellular localization of ZmPCP, the gene was cloned from maize cDNA and fused with GFP in a pCAMBIA1302 vector under the CaMV35S promoter. The recombinant plasmid was transformed into *Agrobacterium tumefaciens* GV3101 and transiently expressed in *Nicotiana benthamiana* leaves. Confocal microscopy revealed ZmPCP-GFP fluorescence specifically enriched at the cell periphery, distinct from the cytoplasmic distribution in GFP controls. Subsequent plasmolysis induced by the 10% NaCl treatment confirmed that the fluorescence remained stably associated with the cell wall (Figure 3), indicating that ZmPCP is synthesized intracellularly and targeted to the apoplastic space via the secretory pathway, consistent with bioinformatic predictions.

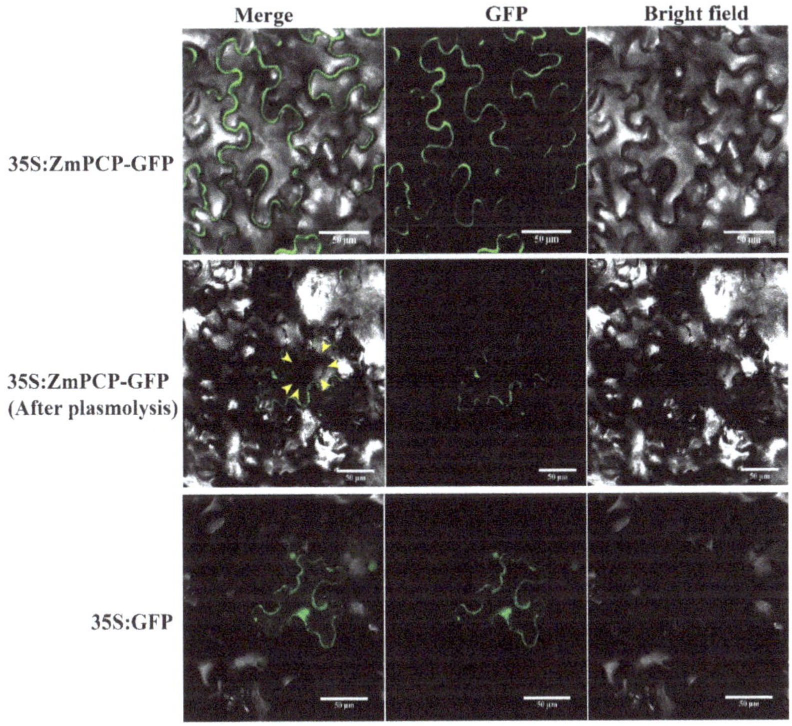

Figure 3. Subcellular localization of ZmPCP in tobacco cells. Scale bar = 20 μm. Note: yellow arrows indicate the position of the plasma membrane after plasmolysis.

2.3. Interaction of ZmPCP with ZmSNAP33

To validate the predicted interaction between ZmPCP and ZmSNAP33 via bioinformatics analysis, this study employed a yeast two-hybrid (Y2H) assay. The preliminary control experiments demonstrated that: (1) the positive control (pGADT7-T + pGBKT7-53) exhibited growth on both SD/-Trp-Leu (double dropout; DDO) and SD/-Trp-Leu-His-Ade (quadruple dropout; QDO) media; (2) the negative control (pGADT7-T + pGBKT7-Lam) showed no growth on QDO medium (Figure 4A), confirming system validity. Subsequent autoactivation and toxicity tests revealed that the pGADT7 + pGBKT7-ZmPCP co-transformants grew on DDO medium (indicating no toxicity) but failed to grow on QDO medium (excluding autoactivation) (Figure 4A). Ultimately, interaction assays confirmed that pGADT7-ZmPCP + pGBKT7-ZmSNAP33 co-transformants grew normally on DDO medium (confirming no toxicity of both proteins) and formed distinct colonies on QDO medium (Figure 4B), demonstrating a specific interaction between ZmPCP and ZmSNAP33.

 https://doi.org/10.3390/plants15050677

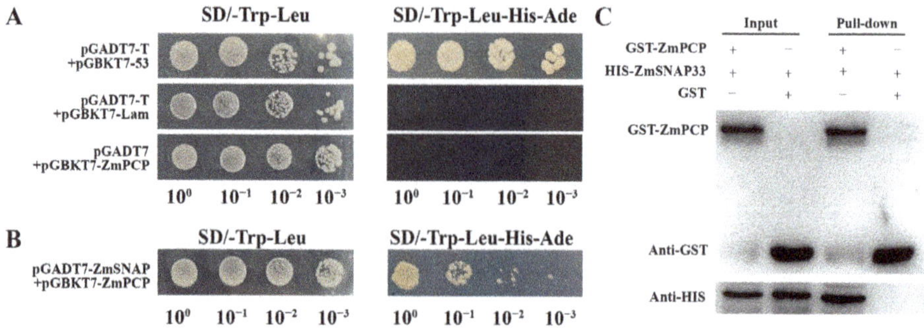

Figure 4. The interactions of ZmPCP with ZmSNAP33. (**A**) Yeast two-hybrid interaction analysis of ZmPCP and ZmSNAP33 with pGADT7-T and pGBKT7-53 as the positive control and pGADT7-T and pGBKT7-lam as the negative control. Validation of ZmPCP toxicity and autoactivation. (**B**) Interaction between ZmPCP and ZmSNAP33. (**C**) GST pull-down assay demonstrating the interaction between ZmPCP and ZmSNAP33. Please refer to Supplementary Figures S1 and S2 for the original images.

Furthermore, GST pull-down assays were employed to substantiate this interaction in vitro. GST (empty vector control) or GST-ZmPCP was incubated with His-ZmSNAP33 and bound to glutathione-agarose beads. After washing, centrifugation, SDS-PAGE, and Western blot analysis, no His-ZmSNAP33 signal was detected in the GST control group, whereas GST-ZmPCP significantly captured His-ZmSNAP33 (Figure 4C), confirming the specific binding. This interaction occurred independently of cofactors, indicating that ZmPCP may directly recognize ZmSNAP33 through specific domains, thereby unambiguously demonstrating their direct interaction in vitro.

2.4. Analysis of the Effects of ZmPCP on Pollen Viability and Drought Resistance

To investigate the effect of *ZmPCP* on pollen viability, the I_2-KI staining method was employed for preliminary detection of untreated B104 (wild type), *ZmPCP*-KO (knockout line), and *ZmPCP*-OE (overexpression line) pollen (Figure 5A). The results show that the pollen viability of B104 and *ZmPCP*-KO ranged from 87% to 93%, whereas that of *ZmPCP*-OE was 83 ± 2.7%, with a significant difference observed (Figure 5B).

In exploring the drought-resistance function of *ZmPCP*, a 28 °C drought treatment revealed that, under untreated conditions, the germination rate of *ZmPCP*-OE was significantly lower than that of the wild-type and knockout lines ($p < 0.05$), indicating that *ZmPCP* overexpression inhibited normal germination. After the 0.5 h treatment, the germination rates of B104 and *ZmPCP*-KO decreased by 22% to 23%, while *ZmPCP*-OE maintained a higher germination rate of 44 ± 5.8%. After the 1 h treatment, this trend continued, with the germination rate of *ZmPCP*-OE (43.5 ± 2.9%) being significantly higher than that of the other lines ($p < 0.01$), demonstrating an early advantage in drought resistance. After the 1.5 h treatment, the germination rate of *ZmPCP*-OE remained significantly higher. By the 2 h treatment, the germination rates of all the lines declined to below 7%, but *ZmPCP*-OE still maintained a relative advantage (Figure 5C,D). In summary, although *ZmPCP* overexpression suppressed the basal germination ability of pollen, it significantly enhanced its tolerance to drought stress.

 https://doi.org/10.3390/plants15050677

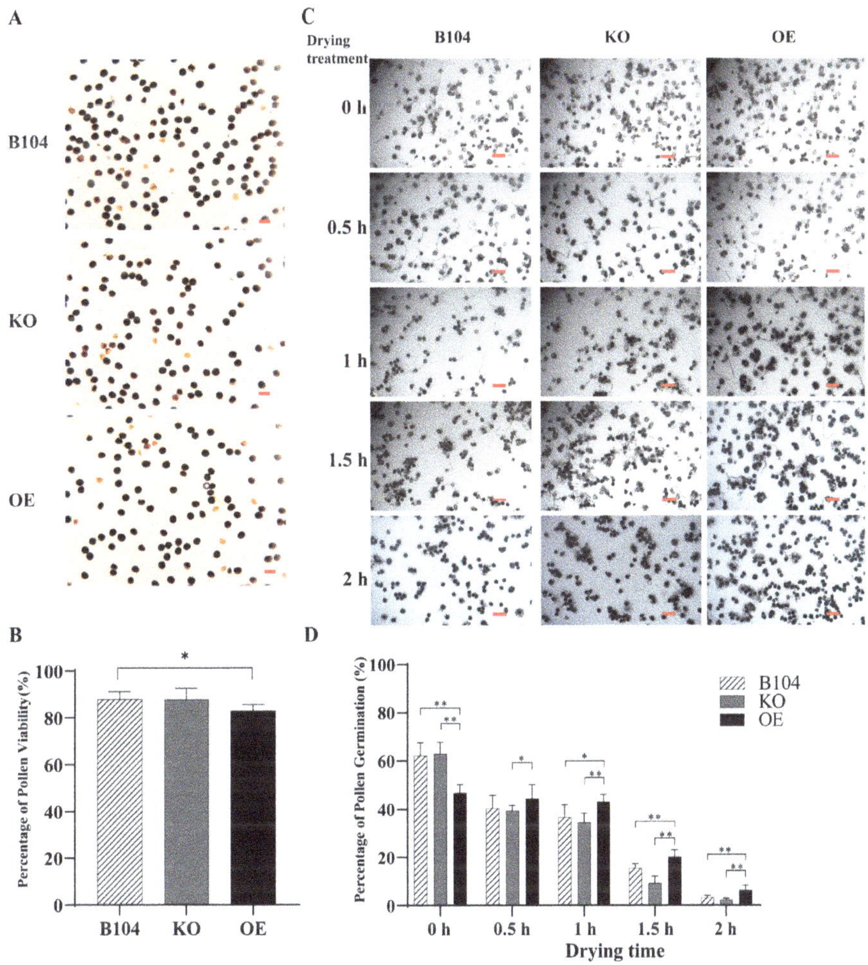

Figure 5. Effect of *ZmPCP* on pollen viability and germination after desiccation treatment (28 °C) for different durations. (**A**) Untreated pollen stained with I_2-KI. Bar = 100 μm. (**B**) Viability statistics of untreated pollen (I_2-KI staining). Bar = 300 μm. The experimental data were obtained from three biological replicates and analyzed for significant differences using Student's *t*-test (* $p < 0.05$). (**C**) Pollen germination images under desiccation treatment (28 °C) at different time points. Please refer to Supplementary Figure S5 for the original images. (**D**) Pollen germination rates under desiccation treatment (28 °C) at different time points. The experimental data were obtained from three biological replicates and analyzed for significant differences using Student's *t*-test (* $p < 0.05$, ** $p < 0.01$).

2.5. *ZmPCP* Promotes Pollen Tube Growth

To investigate the effects of the maize *ZmPCP* gene on pollen tube growth, silk samples from B104 (wild type), *ZmPCP*-KO, and *ZmPCP*-OE were collected at 0.5 h intervals after self-pollination. The maximum pollen tube elongation distance within the field of view was measured from the stigma attachment point (Figure 6A).

https://doi.org/10.3390/plants15050677

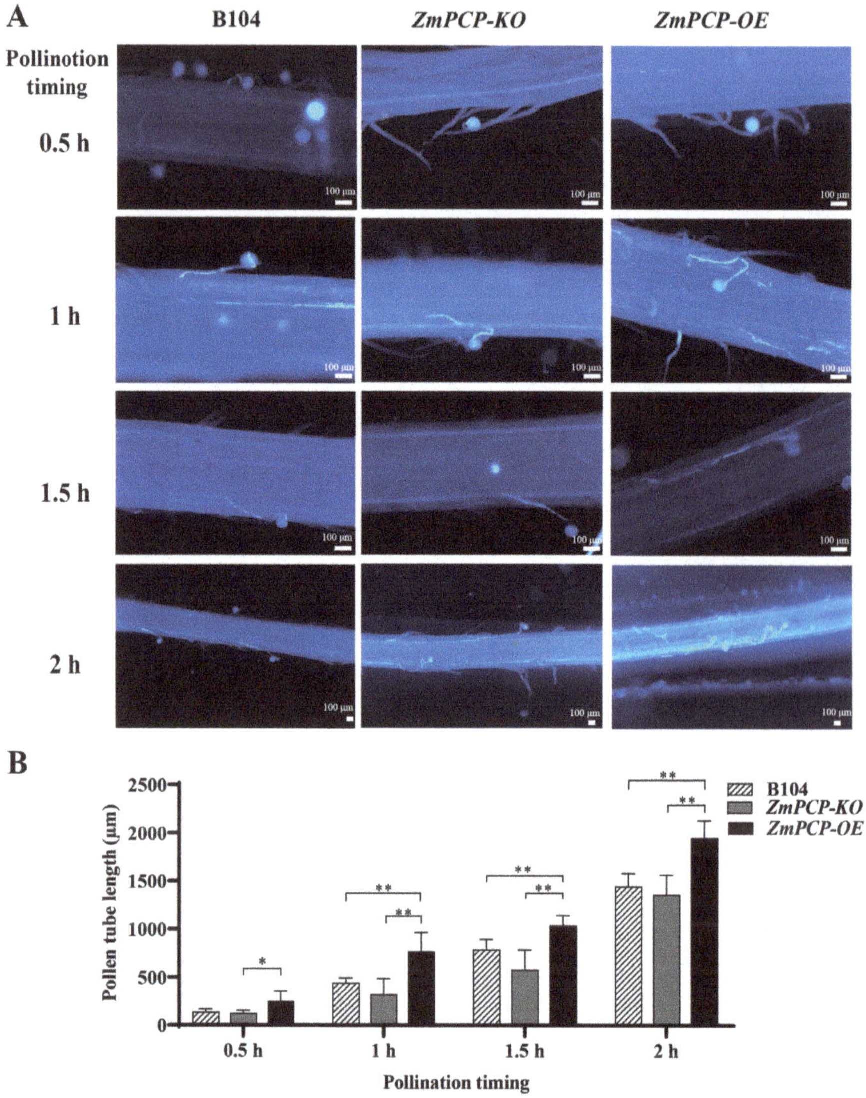

Figure 6. Effect of *ZmPCP* on pollen tube elongation. (**A**) Fluorescence microscopy images showing pollen tube elongation in the styles of B104, *ZmPCP*-KO, and *ZmPCP*-OE lines under self-pollination. Scale bar = 100 μm. (**B**) Quantification of pollen tube length in the styles of B104, *ZmPCP*-KO, and *ZmPCP*-OE lines. The experimental data were obtained from three biological replicates and analyzed for significant differences using Student's *t*-test (* $p < 0.05$, ** $p < 0.01$).

The results demonstrated a significant positive correlation ($p < 0.01$) between the *ZmPCP* expression levels and pollen tube growth kinetics. The B104 pollen tubes elongated progressively from 136 ± 5.3 μm at 0.5 h to 1452 ± 28 μm at 2 h; *ZmPCP*-KO exhibited significantly reduced lengths versus B104 at all time points (e.g., 1363 ± 22 μm at 2 h; $p < 0.05$); *ZmPCP*-OE displayed accelerated growth throughout, reaching 253 ± 9.1 μm (1.86-fold of B104) at 0.5 h and achieving 1956 ± 31 μm at 2 h (1.44-fold longer than *ZmPCP*-KO) (Figure 6B). This dynamic change is consistent with the typical pattern of pollen tube elongation in maize, in which pollen tubes begin to elongate early after pollination and

 https://doi.org/10.3390/plants15050677

show an accelerating trend with time [20]. These findings indicate that *ZmPCP* deficiency significantly suppresses pollen tube elongation ($p < 0.01$), while its overexpression enhances this process, demonstrating that *ZmPCP* acts as a positive regulator of pollen tube growth and dirctly determines elongation velocity.

2.6. RNA-Seq Analysis of Pollen

To investigate the regulatory role of *ZmPCP* in maize pollen development, we performed an RNA-seq analysis on mature pollen from *ZmPCP* overexpression (OE) and knockout (KO) lines under normal growth conditions, with three biological replicates. Applying a threshold of $|\log_2 (Fold\ Change)| > 1$ and adjusted $p < 0.05$, we identified 1664 differentially expressed genes (DEGs), including 673 upregulated and 991 downregulated genes (Figure 7A,B).

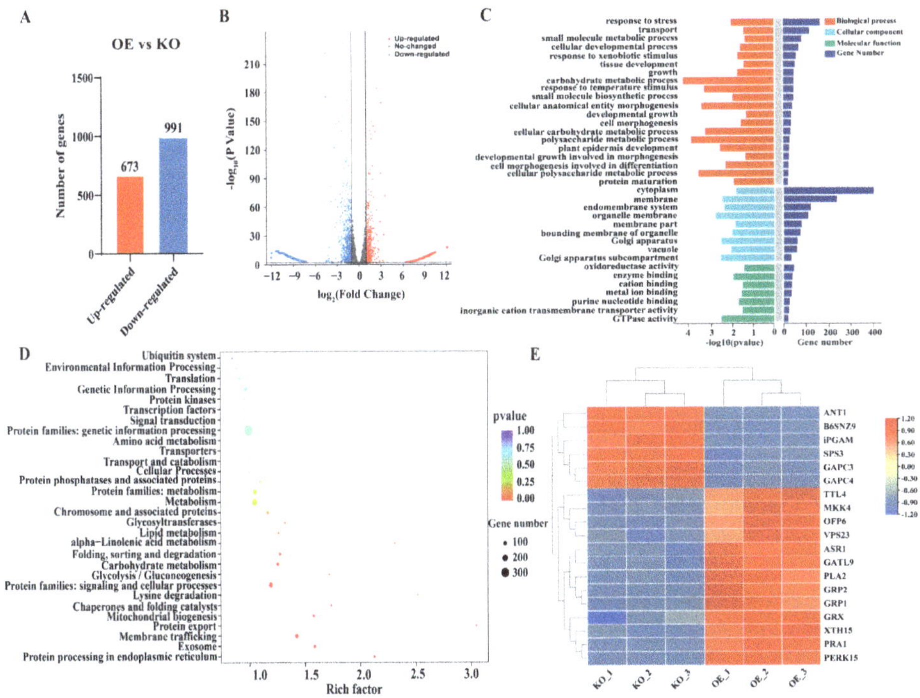

Figure 7. RNA-Seq analysis of *ZmPCP-OE* and *ZmPCP-KO* mature pollen. (**A**) Number of differentially expressed genes. Upregulated and downregulated genes are derived from *ZmPCP-OE* and compared with *ZmPCP-KO*. (**B**) Volcano plot analysis of differentially expressed genes. Blue indicates that the differentially expressed genes are significantly downregulated, while red indicates that they are significantly upregulated. (**C**) GO enrichment analysis of differentially expressed genes. (**D**) KEGG enrichment analysis of differentially expressed genes. (**E**) Heatmap of differentially expressed genes. Blue indicates lower expression levels; red indicates higher expression levels. Dendrogram shows clustering results.

A Gene Ontology (GO) enrichment analysis revealed significant enrichment of pollen DEGs in biological processes related to stress response, transmembrane transport, and cellular development. The subcellular localization predictions (plasma membrane/endomembrane system) suggest that *ZmPCP* may function through the apoplastic pathway. Molecular functions were primarily associated with oxidoreductase activity and ion transport ($p < 1 \times 10^{-5}$), indicating that *ZmPCP* modulates pollen energy supply by

https://doi.org/10.3390/plants15050677

regulating redox homeostasis and nucleotide metabolism (Figure 7C). A KEGG pathway analysis further identified enrichment in 30 key pathways, including carbohydrate, lipid, and amino acid metabolism, as well as genetic information processing, signal transduction, and cellular structure maintenance. These findings suggest that *ZmPCP* coordinates a "metabolism-signaling-structure" mechanism during pollen development.

These results indicate that *ZmPCP* acts as a cross-pathway regulatory hub, coordinating a "metabolism-signaling-structure" trinity mechanism during pollen development. To decipher the molecular mechanisms underlying drought resistance and pollen tube growth dynamics, we systematically analyzed pollen transcriptomes and prioritized functionally characterized DEGs (Table S4). The literature evidence implicates drought-responsive genes *Zm00001eb275850 (GRP2)*, *Zm00001eb388380 (GRP1)*, *Zm00001eb255210 (TTL4)*, *Zm00001eb110700 (PLA2)*, *Zm00001eb351170 (GRX)*, and *Zm00001eb407710 (ASR1)* in stress tolerance [21–26]. It also indicates energy metabolism genes *Zm00001eb252870 (ANT1)*, *Zm00001eb184000 (GAPC3)*, *Zm00001eb246370 (GAPC4)*, *Zm00001eb368990 (iPGAM)*, and *Zm00001eb275240 (SPS3)* in ATP synthesis [27–31]. Further, it outlines vesicle trafficking/cell wall dynamics genes *Zm00001eb290310 (PRA1)*, *Zm00001eb047590 (OFP6)*, *Zm00001eb117750 (PERK15)*, *Zm00001eb259630 (VPS23)*, *Zm00001eb194800 (GATL9)*, and *Zm00001eb312000 (XTH15)* in polar growth [32–37]. Finally, signaling regulators are indicated, including *Zm00001eb257340 (MKK4)*, *Zm00001eb149520 (B6SNZ9)*, and *Zm00001eb286870 (CML15)*, in transduction cascades [38–40].

In the *ZmPCP*-OE line, the expression levels of those genes associated with stress response, pollen tube elongation, cell wall metabolism, and vesicle trafficking were significantly elevated compared to the *ZmPCP-KO* line ($p < 0.05$). This transcriptional profile corresponds to enhanced drought tolerance and accelerated pollen tube growth. Paradoxically, the expression of specific energy metabolism-related genes was downregulated in *ZmPCP*-OE, which may contribute to the observed reduction in the pollen germination rate.

2.7. Overexpression of ZmPCP Inhibits Maize Seed Germination

To investigate the impact of *ZmPCP* on maize seed germination, the germination rates of the B104 (wild type), *ZmPCP*-KO, and *ZmPCP*-OE lines were compared on day 7 of soil-based cultivation. The germination rates were 94% (B104), 93% (*ZmPCP*-KO), and 69% (*ZmPCP*-OE), respectively (Figure S4), with *ZmPCP*-OE exhibiting significantly reduced germination compared to the wild-type and knockout lines ($p < 0.05$). These results indicate that *ZmPCP* overexpression suppresses maize seed germination and functions as a negative regulator in this process.

2.8. Promotion of Germination and Seedling Growth in Arabidopsis Under Stress Through ZmPCP Overexpression

To investigate the functional role of *ZmPCP* in stress responses, the recombinant vector pCAMBIA1302-*ZmPCP* was transformed into Arabidopsis. Transgenic lines were selected via antibiotic resistance screening and systematically validated using molecular biology approaches. A genomic PCR analysis (Figure 8A) confirmed the successful integration of *ZmPCP* into the Arabidopsis genome, with the wild type (WT) serving as the negative control; RT-qPCR assays (Figure 8B) demonstrated significantly elevated *ZmPCP* transcript levels in the OE3 and OE6 lines, meeting the overexpression criteria [41], with OE3 and OE6 exhibiting relatively smaller variance. Homozygous T3 lines were selected for functional assays.

Under simulated abiotic stress conditions, *ZmPCP* overexpression markedly altered the seed germination dynamics. In the standard MS medium, both overexpression lines (OE3 and OE6) and WT exhibited >90% germination at 24 h, reaching 100% by 48 h. Under 300 mM mannitol stress, the germination rates at 24 h were significantly reduced (OE3:

 https://doi.org/10.3390/plants15050677

29% ± 3.1; OE6: 21% ± 2.8 vs. WT: 35% ± 3.5; $p < 0.05$), with completion delayed to 72 h (Figure 8C). The ABA treatment similarly suppressed germination, with the overexpression lines consistently lagging behind the WT during 24–48 h. Notably, under the 2 μM ABA treatment, OE6 exhibited a more obvious reduction in germination rate than OE3, which may be due to small changes in transgenic insertion sites and genetic backgrounds leading to phenotypic differences that are not strictly proportional to expression level. These results indicate that *ZmPCP* modulates germination timing in response to osmotic stress and ABA, thereby participating in seed germination regulation.

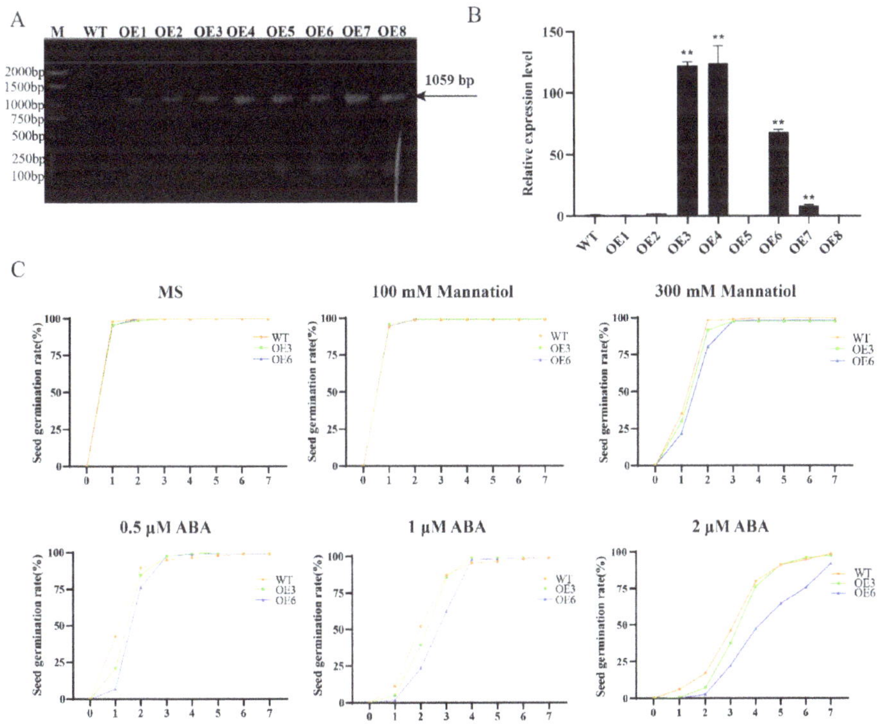

Figure 8. The effects of *ZmPCP* overexpression on seed germination. (**A**) PCR verification of *ZmPCP* in transgenic *Arabidopsis* lines. Please refer to Supplementary Figure S3 for the original image. (**B**) Validation of *ZmPCP* expression levels in transgenic *Arabidopsis*. The experimental data were obtained from three biological replicates and analyzed for significant differences using Student's *t*-test (** $p < 0.01$). (**C**) Germination rate of *Arabidopsis thaliana* cultured for 7 days on MS solid medium containing different concentrations of mannitol and ABA.

Further phenotypic analysis of the seedlings revealed no significant differences in root length among WT, OE3, and OE6 under control conditions (Figure 9A). At low ABA concentrations (0.5–1 μM), the root lengths of the overexpression lines were marginally greater than WT (not statistically significant). At 2 μM ABA, the OE3 and OE6 roots (14 ± 2.1 mm and 12.9 ± 1.4 mm, respectively) exceeded WT (10.6 ± 1.5 mm) by 21.7–32.1% ($p < 0.01$) (Figure 9B). This demonstrates that *ZmPCP* overexpression enhances root adaptation to high-concentration ABA stress, potentially through modulation of stress signaling pathways.

https://doi.org/10.3390/plants15050677

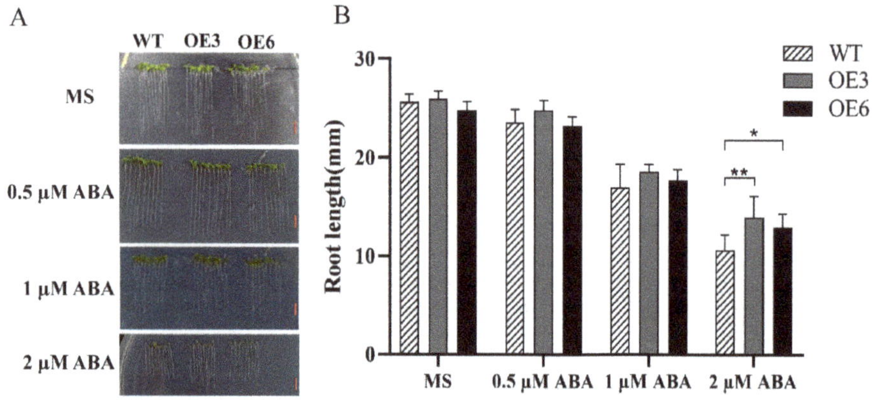

Figure 9. Roots of transgenic Arabidopsis under ABA treatment. (**A**) Roots grown for 7 days under different concentrations of ABA, with a scale bar of 5 mm. Please refer to Supplementary Figure S6 for the original images. (**B**) Data of root lengths of plants grown for 7 days under different concentrations of ABA. The experimental data were obtained from three biological replicates and analyzed for significant differences using Student's *t*-test (* $p < 0.05$, ** $p < 0.01$).

3. Discussion

Papain-like cysteine proteases (PLCPs) serve as crucial regulators in plants, extensively participating in developmental processes and stress responses. Functional studies on identified PLCP members in maize (such as Mir1 and CP1B) remain limited [42–44]. This study reveals that ZmPCP (belonging to the THI1 clade of the C1A subfamily) significantly enhances maize drought resistance by improving cellular osmotic adjustment and membrane stability and additionally regulates pollen germination [17,45]. The protein encoded by this gene is an apoplastic secretory protein possessing a conserved Peptidase_C1 domain, exhibiting high homology with its ortholog in *Zea mays* ssp. *parviglumis* (*Mexican teosinte*), suggesting evolutionary conservation (Figure 1). Subcellular localization confirmed ZmPCP as an apoplastic secretory protein (Figure 3), consistent with the results from immunogold electron microscopy in pollen [46]. Notably, *ZmPCP* overexpression alters pollen pore morphology, indicating its potential involvement in cell wall remodeling (e.g., pectin degradation) through extracellular proteolysis. This mechanism is analogous to that of Arabidopsis *CP1*, which cleaves extensins to promote polar growth [47].

Furthermore, this study identified an interaction between ZmPCP and the membrane-localized SNARE protein ZmSNAP33 (Figure 5). SNARE (soluble N-ethylmaleimide-sensitive-factor attachment protein receptor) family proteins were initially discovered in yeast. Based on cluster analyses of their SNARE domain amino acid sequences, SNAREs are classified into four subgroups: Qa-SNAREs, Qb-SNAREs, Qc-SNAREs, and R-SNAREs. They are also categorized as vesicle-associated SNAREs (v-SNAREs) or target membrane SNAREs (t-SNAREs) [48]. SNARE proteins play essential roles in vesicle trafficking, immune secretion, exocytosis during pollen tube growth, and the fusion of secretory vesicles with the plasma membrane at the pollen tube apex [20]. SNAP25-type SNAREs, which are located on the plasma membrane, function as key components of the SNARE complex by assembling with a syntaxin (Qa-SNARE) and a vesicle-anchored R-SNARE to mediate membrane fusion [49]. *ZmSNAP33* encodes a SNAP25-type SNARE protein, and its Arabidopsis homolog *AtSNAP33* has been reported to cause delayed pollen hydration on mutant stigmas [50]. One of the cellular responses in the stigma associated with pollen hydration involves vesicle trafficking in the stigmatic papilla, which is presumed to deliver cargo that facilitates water release to the pollen grain [51]. Moreover, OCP, featuring cys-

 https://doi.org/10.3390/plants15050677

teine protease activity and whose loss of function leads to abnormal pollen development, interacts with OsRACK1A or OsSNAP32 physically, in vitro and in vivo, and can suppress the expression of OsSNAP32 [52]. Therefore, through its interaction with ZmSNAP33, ZmPCP may regulate the vesicle trafficking in the stigmatic papilla during the process of pollen hydration (Figure 10). In addition, several W-boxes (WRKY transcription factor binding sites, TTGAC), MYB binding sites (WAACCA), and ABREs (abscisic acid responsive elements, AAACGTGA) have been identified in the promoter region of *OsSNAP32*. These cis elements suggest that *OsSNAP32* participates in various signal transduction pathways related to plant responses to biotic and abiotic stresses [49]. Accordingly, its maize homolog ZmSNAP33 may have a pleiotropic role in responding to environmental stressors, potentially mediated through its interaction with ZmPCP (Figure 10).

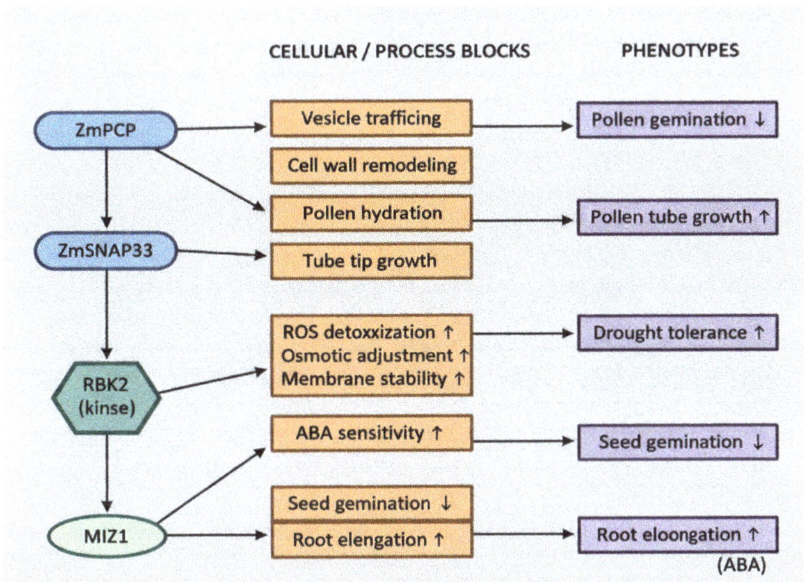

Figure 10. The possible functional scheme of ZmPCP. ZmSNAP33: a SNAP25-type SNARE protein; SNARE: soluble N-ethylmaleimide-sensitive-factor attachment protein receptor; RBK2: a pathogen-induced receptor-like cytosolic kinase 2; MIZ1: MIZU-KUSSEI 1. Upward arrows (↑) denote an increase or enhancement of the corresponding cellular process or phenotype, whereas downward arrows (↓) indicate a decrease or reduction in the related process or phenotype.

Pollen fertility is a critical factor for successful plant sexual reproduction and is regulated by diverse genetic, environmental, and physiological–biochemical factors. Our results demonstrate that *ZmPCP* plays a crucial regulatory role within this process. Overexpression of *ZmPCP* reduced the germination rate of non-desiccated pollen, potentially due to disrupted energy metabolism or aberrant cell wall remodeling. Several genes associated with energy metabolism were significantly upregulated in *ZmPCP*-KO pollen compared with *ZmPCP*-OE (e.g., *Zm00001eb184000 (GPC3)*, *Zm00001eb246370 (GPC4)*, *Zm00001eb149730 (PGAM)*, and *Zm00001eb368990 (PGAM)* in Table S2). This pattern indicates that *ZmPCP* overexpression may restrict energy metabolism during early pollen development, potentially contributing to reduced pollen activity. The mutations of *NtiPGAM* (2, 3-bisphosphoglycerate-independent phosphoglycerate mutase) impaired glycolysis, limited the energy supply, and ultimately led to defective pollen development and pollen tube growth in the mutant plants [53]. These parallels suggest that *ZmPCP* influences pollen

 https://doi.org/10.3390/plants15050677

performance in part through modulation of carbohydrate metabolic pathways. Despite its negative effect on germination, *ZmPCP* overexpression markedly enhanced drought tolerance after a 1 h desiccation treatment, and drought resistance was significantly enhanced, a mechanism involving the accumulation of osmolytes (e.g., sugars) and upregulation of antioxidant enzyme activity [54]. Additional transcriptomic evidence supports this mechanism: several stress-response genes were upregulated in *ZmPCP*-OE pollen, including a pathogen-induced receptor-like cytosolic kinase *RBK2 (Zm00001eb165930)*, a member of the Arabidopsis RLCK-VIb family known to enhance ROS detoxification and improve drought and salt tolerance [55,56]. Meanwhile, *MIZU-KUSSEI 1 (MIZ1) (Zm00001eb069080)* was also upregulated, which may play a role in drought stress [57]. *AtMIZ1* was also shown to improve drought avoidance and increase root cell viability under hydro-stimulated conditions in Arabidopsis [58]. Notably, although *ZmPCP* overexpression inhibited germination, it promoted pollen tube elongation, with its expression level positively correlated with pollen tube growth rate. Transcriptome profiling further supports a role for *ZmPCP* in regulating cell wall synthesis: the genes involved in cell wall synthesis were downregulated in *ZmPCP*-KO compared with *ZmPCP*-OE pollen (e.g., *Zm00001eb275850 (GRP2)*, *Zm00001eb388380 (GRP1)*, *Zm00001eb303890 (RALF)*, and *Zm00001eb117750 (PERK15)* in Table S3). This is consistent with observations in Arabidopsis, where the *perk5-1 perk12-1* double mutant exhibits shorter pollen tubes, reduced pollen transmission, and decreased seed set due to altered cell wall polysaccharide composition and disrupted ROS homeostasis [34]. Together, these findings suggest that *ZmPCP* promotes pollen tube elongation by coordinating cell wall synthesis, modification, and redox signaling at the growing tip (Figure 10).

A pollen transcriptome analysis further revealed that differentially expressed genes were predominantly enriched in maize stress responses, growth and development, carbohydrate metabolic processes, signal transduction, and protein secretion capacity. ZmPCP likely functions as a cross-pathway regulatory node, exerting significant control over pollen germination and pollen tube growth by modulating the expression of genes associated with stress responses, energy metabolism, vesicle trafficking, cell wall structure, and signal transduction (Figure 10).

During seed germination, *ZmPCP* overexpression significantly reduced maize seed germination rates, contrasting with the promotive role of *HvPap-1* [59]. ABA inhibits seed germination, while ABA-deficient mutants germinate more rapidly than the wild type under the same conditions [60]. The reduced germination of *ZmPCP*-overexpressing seeds suggests the possible involvement of ABA-associated regulatory pathways. Consistent with this, under stress treatments with mannitol and ABA, germination of *ZmPCP*-overexpressing Arabidopsis seeds exhibited a certain delay but did not affect overall germination (Figure 8). Previous reports have shown that ABA can regulate the expression of *MIZ1* [57]. Therefore, it is possible that *ZmPCP* influences ABA-associated processes. Under high-concentration ABA treatment, the root length of *ZmPCP*-overexpressing Arabidopsis lines was significantly longer than that of the wild type, indicating that *ZmPCP* coordinately regulates germination and root growth through the ABA signaling pathway [9,61]. The enhanced ABA responsiveness may be related to the *ZmPCP*-induced upregulation of *RBK2 (Zm00001eb165930)*, a kinase associated with cell wall biosynthesis and ROS-related signaling processes [55]. Notably, PLCP family members (e.g., *RD21*) have been demonstrated to enhance plant resistance by degrading viral pathogenicity factors [62], suggesting that *ZmPCP* may possess analogous immune-regulatory potential.

https://doi.org/10.3390/plants15050677

4. Materials and Methods

4.1. Plant Materials and Growth Conditions

To analyze the effects of *ZmPCP* (gene ID: LOC100280441) on seed germination of maize, the seeds of maize inbred lines B104, *ZmPCP*-OE and *ZmPCP*-KO (generated in our previous work) [17] were germinated in soil mixture composed of peat soil and vermiculite thoroughly mixed at a 1:1 ratio and watered every 3 d. The seedlings were grown in plastic pots (7 cm top diameter, 5 cm bottom diameter, and 7.5 cm height), with five plants per pot, under controlled conditions of 27/22 °C (day/night) and 60% relative humidity with a 14/10 h photoperiod and a light intensity of 2000 Lux. Count the germination rate of maize plants when they grow to the seventh day. All *Arabidopsis thaliana* plants and *Nicotiana benthamiana,* grown in the same soil mixture and in pots of the same specifications, were maintained under a 16 h light/8 h dark cycle with approximately 60% relative humidity at 22 °C in a plant growth chamber. All experiments were conducted with at least three independent biological replicates.

4.2. Bioinformatics Analysis of ZmPCP

Genomic data and annotation files for maize (B73) and Mexican teosinte (*Zea mays* ssp. *mexicana)* were obtained from the NCBI database (https://www.ncbi.nlm.nih.gov/ (accessed on 30 January 2024)) and ZEAMAP (https://db.cngb.org/zeamap/ (accessed on 30 January 2024)), respectively. Arabidopsis thaliana cysteine protease gene sequences were acquired from the TAIR database (https://www.arabidopsis.org/ (accessed on 30 January 2024)). Genomic data extraction and conversion to protein sequences were performed using TBtools (version 2.323) [63]. Candidate family members were identified via bidirectional Blast alignment (E-value $\leq 1 \times 10^{-5}$) against published Arabidopsis cysteine protease protein sequences as Query Seqs. Redundant sequences were curated through motif analysis (MEME Suite: https://meme-suite.org/meme/ (accessed on 30 January 2024)), domain validation (NCBI CDD: https://www.ncbi.nlm.nih.gov/Structure/bwrpsb/bwrpsb.cgi (accessed on 30 January 2024)), and InterPro structural analysis (https://www.ebi.ac.uk/interpro/ (accessed on 30 January 2024)). Multiple sequence alignment was conducted using MUSCLE in MEGA 11 software (default parameters), followed by neighbor-joining (NJ) phylogenetic tree construction (Bootstrap 1000–1500 replicates; Jones–Taylor–Thornton model). The evolutionary tree was visually refined using iTOL (version 7) (https://itol.embl.de/ (accessed on 30 January 2024)). Protein tertiary structures of ZmPCP homologs (Mexican teosinte and Arabidopsis) were predicted via AlphaFold2 to analyze evolutionary relationships and structural divergence.

4.3. RNA Extraction and Quantitative Real-Time PCR (qRT-PCR)

Referring to the method of Li [17], the main steps are as follows: Total RNA in maize tissues was extracted using RNA-Solv® Reagent (Omega Bio-Tek, Norcross, GA, USA), and a 2 μg RNA sample was used to synthesize the first-strand cDNA via the 5 × All-In-One MasterMix with AccuRT Genomic DNA Removal Kit (Applied Biological Materials Inc., Richmond, BC, Canada). The qRT-PCR was carried out using a Hieff qPCR SYBR Green Master Mix kit (YEASEN Biotechnology, Shanghai, China), and the gene-specific primers of PCP can be referred to in the previous research of our research group [17]. The quantification method $2^{-\Delta\Delta CT}$ was used to assess the relative expression level of PCP after normalization based on *ZmUBI* expression [64]. Data were represented as relative expression (mean ± SD) from three biological replicates.

https://doi.org/10.3390/plants15050677

4.4. Subcellular Localization

Nicotiana benthamiana was selected due to its high efficiency in *Agrobacterium*-mediated transient expression, enabling rapid assessment of protein subcellular localization [65]. To determine the subcellular localization of ZmPCP, the CDS region of ZmPCP was cloned downstream of CaMV 35S promoter and into the pCAMBIA1302-GFP vector to construct the 35S: *ZmPCP*-GFP expression vector. The recombinant vector was transformed into *Agrobacterium* GV3101 and infiltrated into tobacco leaf epidermal cells using the *Agrobacterium*-mediated method, incubating at 28 °C for 2 d. Fluorescence signals were observed using a confocal laser microscope (Nikon, Tokyo, Japan).

4.5. Identification of ZmSNAP33 as a Candidate Interacting Partner of ZmPCP and Yeast Two-Hybrid Assay

To identify potential interacting partners of ZmPCP, we first referred to previous studies reporting that OsCP (an oryzain alpha-chain precursor featuring cysteine protease activity) interacts with OsSNAP32 (synaptosome-associated protein of 32 kD) physically in vitro and in vivo in rice [52]. Using the OsSNAP32 protein sequence as a query, we performed BLASTP searches against the maize proteome and identified several SNAP25 homologs, including protein SNAP33 (GeneID:100217287). These ZmSNAP33 candidates were subsequently cloned and subjected to yeast two-hybrid screening with ZmPCP. Through this stepwise validation, the ZmSNAP33 protein reported in this study was identified as the interacting partner of ZmPCP.

The interactions of ZmPCP and ZmSNAP33 were studied using the Gal4-based yeast two-hybrid (Y2H) system, following the method of Liang [66] and the manufacturer's instructions (Zoman Bio., Beijing, China). The CDS of *ZmPCP* was cloned into the decoy vector (pGBKT7) to generate the bait vector pGBKT7-ZmPCP, and the CDSs of ZmSNAP33 were cloned into pGADT7 as prey. pGBKT7-53/pGADT7-T and pGBKT7-Lam/pGADT7-T were used as positive and negative controls, respectively. After co-transformation into Y2HGold competent cells (Zoman Bio., Beijing, China), transformants were first selected on SD/-Leu/-Trp medium, and interaction was tested on SD/-Leu/-Trp/-His/-Ade medium. One ZmSNAP33 candidate showing consistent growth on selective medium was identified as the interacting partner of ZmPCP. Primers used for vector construction are listed in Table S1.

4.6. Prokaryotic Protein-Induced Expression and Protein Purification

The coding sequences (CDSs) of ZmPCP and ZmSNAP33 were cloned into the pGEX4T-1 and pET-30a (+) vectors, respectively, using NcoI and SpeI for digestion and subsequent ligation. The constructed vectors were then transformed into Escherichia coli BL21(DE3) cells. Clones were selected using kanamycin (50 μg/mL) as a selection marker. *E. coli* cells harboring the genes of interest were grown overnight at 37 °C with shaking at 200 rpm in 5 mL of LB liquid medium supplemented with 50 μg/mL kanamycin. Subsequently, 1 mL of this saturated culture was inoculated into 50 mL of fresh LB medium containing 50 μg/mL kanamycin and cultured at 37 °C until the optical density at 600 nm (OD600) reached 0.6. The incubation temperature was then lowered to 16 °C, and protein expression was induced with 0.1 mM isopropyl-β-D-thiogalactopyranoside (IPTG) for 14 h. Purification of ZmPCP-GST was performed using the GST-tag Protein Purification Kit (Beyotime Biotechnology, Shanghai, China) following the manufacturer's instructions, while ZmSNAP33-His was purified using a His-tagged protein purification kit (CWBIO, Taizhou, China) as per the manufacturer's protocol. Finally, 12% SDS-PAGE electrophoresis was used to analyze the expression product and the purified protein.

 https://doi.org/10.3390/plants15050677

4.7. GST –Pull-Down and Western Blot

To confirm the direct interaction of ZmPCP with ZmSNAP33, we performed a GST pull-down assay. ZmPCP-GST or GST proteins were incubated with ZmSNAP33-His protein in 100 μL of binding buffer for 2 h at 4 °C with continuous rotation [67]. After incubation, 30 μL of pre-equilibrated GST agarose gel was added, followed by incubation at 4 °C with gentle agitation (40 rpm) for 16 h (overnight). A 20 μL sample was taken as input, while the remaining samples were washed thoroughly and subjected to SDS-PAGE. For the Western blot analysis, proteins separated by SDS-PAGE were transferred to PVDF membranes. After blocking the membrane, it was incubated with the primary antibody (diluted 1:2000), followed by the secondary antibody (diluted 1:5000). After TBST washing, protein bands were visualized using ECL substrate and imaged with a chemiluminescence detection system. Band intensities were quantified using ImageJ software (v1.8.0) with normalization to input controls. The primers used in this experiment are listed in Supplemental Table S1.

4.8. Pollen Viability and Germination in Vitro

Pollen viability was examined with I_2-KI staining methods [68]. Pollen grains that were stained blue after 8 min of staining in 0.5% I_2-KI solution were also counted as viable. The proportion of about 50 pollens with different staining depth was counted under the microscope (Eclipse Ni-U, Nikon, Tokyo, Japan) to evaluate the maturity or vitality of the population, and photos were taken for preservation [17]. In vitro germination experiments were carried out. The collected pollens were placed in disposable Petri dishes. After drying at 28 °C for 0, 0.5, 1, 1.5 and 2 h, an appropriate amount of pollen was spread on 1 × 1 cm pollen germination solid medium (the components included 100 mg/L Ca $(NO_3)_2$, 10 mg/L H_3BO_3, 10 mg/L $MgCl_2$, 15% *w*/*v* sucrose and 1.0% *w*/*v* agar). After incubation at 28 °C for 4 h, the germination of no less than 200 pollen particles in each field was counted and photographed with a microscope (Eclipse Ni-U, Nikon, Tokyo, Japan). A pollen grain was considered germinated when its tube length reached pollen diameter [69]. All viability and germination tests were conducted in three biological replications.

4.9. Aniline Blue Staining of Pollen Tubes

Pollen collected from maize plants was evenly distributed to self-pollinated silks, and silk samples were collected at 0.5 h, 1 h, 1.5 h and 2 h after pollination. The samples were fixed in Carnoy stationary solution (absolute ethanol: chloroform: glacial acetic acid mixed at 6:3:1 volume ratio) for 24 h. After fixation, the silks were rinsed in deionized water and then softened in 4 m KOH solution for 4 h, followed by another rinse with deionized water. The softened silks were then incubated in aniline blue staining solution (0.1 M phosphate buffer containing 0.1% aniline blue, pH = 7.0) for 12 h. After staining, the silks were washed with deionized water, cut into 1.5 cm segments, mounted in 50% glycerol, and examined under the ultraviolet filter of fluorescence microscope (Eclipse Ni-U, Nikon, Tokyo, Japan) with near-UV excitation [70].

4.10. Transcriptome Analysis of Pollen

In this study, mature pollen from *ZmPCP-OE* and *ZmPCP*-KO maize plants grown under normal conditions was used for transcriptome sequencing. Three biological replicates were prepared for each genotype. Total RNA per sample was used for cDNA library construction and Illumina sequencing (BGI Genomics Co. Ltd., Beijing, China), and RNA-seq data were processed, assembled, and annotated. Bioinformatic analysis was performed using TBtools software. First, clean data were aligned to the reference genome using the Hisat2 plugin. Then, read counts for each gene were quantified with StringTie. The differentially

https://doi.org/10.3390/plants15050677

expressed genes (DEGs) were conducted using the DESeq2 plugin (version 2.323) to compute the fold change and *p*-value for each gene. Genes satisfying $|\log_2 \text{(fold change)}| >= 1$ and *p*-value < 0.05 were identified as differentially expressed. Finally, TBtools plugins were utilized to generate a volcano plot of differentially expressed genes and to perform both GO and KEGG enrichment analyses.

4.11. Heterologous Expression of *ZmPCP* in *Arabidopsis thaliana*

Arabidopsis is a preferred model for physiological, biochemical, genetic, and molecular studies due to its compact and well-annotated genome, ease of cultivation and manipulation, short life cycle, and high seed yield [71].

The CDS of *ZmPCP* was inserted into the pCAMBIA1302 overexpression vector (Takara Bio, Inc. Shiga, Japan). *Agrobacterium tumefaciens* GV3101 (Zoman, Beijing, China) harboring the vector p pCAMBIA1302-*ZmPCP* under 35S promoter was transformed into *Arabidopsis* using the floral dip method. The positive lines were selected using 50 mg/L hygromycin for further analysis. The T2 plants that showed 100% resistance to hygromycin were considered homozygous transformants in the T3 generation, and then the T3 homozygous transgenic lines OE-1 and OE-2 with high expression levels of *ZmPCP* were used for further analysis as the experimental materials.

4.12. Germination Assessment

A total of 63 wild-type (WT) and transgenic (*ZmPCP*-overexpressing lines OE-3 and OE-6) *Arabidopsis thaliana* seeds were surface-sterilized and planted on the Murashige and Skoog (MS) medium supplemented with mannitol (0, 100, or 300 mM) or ABA (0, 0.5, 1, or 2 μM) in triplicate. All the seeds were planted in a tissue culture incubator under a 16/8 h day/night cycle at 22 °C for germination and vertically cultivated after 4 days. The germination (emergence of radicles) was monitored daily. Subsequently, the root lengths of seedlings treated with 0.5, 1, or 2 μM ABA were measured and photographed on days 7 of vertical cultivation.

4.13. Statistical Analysis

All statistical data were calculated from three replications using DPS 8.0 software package and are reported as means ± standard error (SE). Differences between different plant lines or treatments were considered significant if $p < 0.05$.

5. Conclusions

In summary, ZmPCP, functioning as an extracellular protease, regulates cell wall remodeling and drought resistance via ZmSNAP33-mediated vesicle trafficking. *ZmPCP*-overexpressing maize exhibited significantly lower pollen viability and seed germination rates compared to wild-type and knockout lines. Nevertheless, *ZmPCP* overexpression conferred significantly enhanced tolerance to drought stress (desiccation treatment) in pollen, mitigating the reduction in pollen viability. Accelerated pollen tube growth was observed in *ZmPCP*-overexpressing maize plants under self-pollination compared to wild-type and knockout lines. However, the upstream regulators of ZmPCP, its downstream target proteins, and the underlying pollination recognition mechanism require further elucidation. This study provides a theoretical foundation and genetic resources for breeding maize with improved drought tolerance and high pollen viability while also offering new insights into the mechanisms of cysteine proteases.

Supplementary Materials: The following supporting information can be downloaded at: https://www.mdpi.com/article/10.3390/plants15050677/s1, Figure S1: The original image of anti GST-ZmPCP; Figure S2: The original image of anti-HIS; Figure S3: The original image of PCR verifica-

https://doi.org/10.3390/plants15050677

tion of *ZmPCP* in transgenic Arabidopsis lines; Figure S4: Germination rates of B104, *ZmPCP*-KO and *ZmPCP*-OE maize seed lines; Figure S5: The original image of viability statistics of untreated pollen (I_2-KI staining); FigureS6: The original image of roots grown for 7 days under different concentrations of ABA; Table S1: Primer sequences; Table S2: RNA-seq analysis showed the expression levels of up-regulated genes in differential genes; Table S3: RNA-seq analysis showed down-regulated gene expression levels in differential genes; Table S4: Reports related to the genes with significantly differential expression in RNA-seq.

Author Contributions: H.L. and W.W. (Wei Wang) designed and supervised the project. W.W. (Wenkang Wang) performed the experiments and bioinformatics analysis. Y.L. wrote the manuscript and prepared the figures and tables and revised the manuscript. All authors have read and agreed to the published version of the manuscript.

Funding: This research was supported by the National Natural Science Foundation of China (No. 31771700 and No. 32272026 to W.W.).

Data Availability Statement: All data generated or analyzed during this study are included in the article and its Supplementary Materials.

Conflicts of Interest: The authors declare no conflict of interest.

References

1. National Bureau of Statistics of China. National Economy Witnessed Steady Progress amidst Stability with Major Development Targets Achieved Successfully in 2024. 2025. Available online: https://www.stats.gov.cn/english/PressRelease/202501/t20250117_1958330.html (accessed on 12 August 2025).
2. Wu, W.; Yue, W.; Bi, J.; Zhang, L.; Xu, D.; Peng, C.; Chen, X.; Wang, S. Influence of climatic variables on maize grain yield and its components by adjusting the sowing date. *Front. Plant Sci.* **2024**, *15*, 1411009. [CrossRef] [PubMed]
3. Li, H.; Tiwari, M.; Tang, Y.; Wang, L.; Yang, S.; Long, H.; Guo, J.; Wang, Y.; Wang, H.; Yang, Q.; et al. Metabolomic and transcriptomic analyses reveal that sucrose synthase regulates maize pollen viability under heat and drought stress. *Ecotoxicol. Environ. Saf.* **2022**, *246*, 114191. [CrossRef] [PubMed]
4. Li, H.; Tang, Y.; Meng, F.; Zhou, W.; Liang, W.; Yang, J.; Wang, Y.; Wang, H.; Guo, J.; Yang, Q. Transcriptome and metabolite reveal the inhibition induced by combined heat and drought stress on the viability of silk and pollen in summer maize. *Ind. Crops Prod.* **2025**, *226*, 120720. [CrossRef]
5. Zhou, L.Z.; Wang, L.; Chen, X.; Ge, Z.; Mergner, J.; Li, X.; Küster, B.; Längst, G.; Qu, L.J.; Dresselhaus, T. The RALF signaling pathway regulates cell wall integrity during pollen tube growth in maize. *Plant Cell* **2024**, *36*, 1673–1696. [CrossRef]
6. Zhong, S.; Li, L.; Wang, Z.; Ge, Z.; Li, Q.; Bleckmann, A.; Wang, J.; Song, Z.; Shi, Y.; Liu, T.; et al. RALF peptide signaling controls the polytubey block in Arabidopsis. *Science* **2022**, *375*, 290–296. [CrossRef]
7. Zhang, D.; Liu, D.; Lv, X.; Wang, Y.; Xun, Z.; Liu, Z.; Li, F.; Lu, H. The cysteine protease CEP1, a key executor involved in tapetal programmed cell death, regulates pollen development in Arabidopsis. *Plant Cell* **2014**, *26*, 2939–2961. [CrossRef]
8. Szewińska, J.; Simińska, J.; Bielawski, W. The roles of cysteine proteases and phytocystatins in development and germination of cereal seeds. *J. Plant Physiol.* **2016**, *207*, 10–21. [CrossRef]
9. Liu, H.; Hu, M.; Wang, Q.; Cheng, L.; Zhang, Z. Role of Papain-Like Cysteine Proteases in Plant Development. *Front. Plant Sci.* **2018**, *9*, 1717. [CrossRef]
10. Hu, M.; Li, Y.; Zhang, X.; Song, W.; Jin, W.; Huang, W.; Zhao, H. Maize sterility gene DRP1 encodes a desiccation-related protein that is critical for Ubisch bodies and pollen exine development. *J. Exp. Bot.* **2022**, *73*, 6800–6815. [CrossRef]
11. Tabuchi, A.; Li, L.C.; Cosgrove, D.J. Matrix solubilization and cell wall weakening by β-expansin (group-1 allergen) from maize pollen. *Plant J.* **2011**, *68*, 546–559. [CrossRef]
12. Gong, F.; Wu, X.; Wang, W. Diversity and function of maize pollen coat proteins: From biochemistry to proteomics. *Front. Plant Sci.* **2015**, *6*, 199. [CrossRef] [PubMed]
13. Suen, D.F.; Huang, A.H. Maize pollen coat xylanase facilitates pollen tube penetration into silk during sexual reproduction. *J. Biol. Chem.* **2007**, *282*, 625–636. [CrossRef] [PubMed]
14. Yang, Y.; Dong, C.; Yu, J.; Shi, L.; Tong, C.; Li, Z.; Huang, J.; Liu, S. Cysteine Protease 51 (CP51), an anther-specific cysteine protease gene, is essential for pollen exine formation in Arabidopsis. *Plant Cell Tissue Organ Cult. (PCTOC)* **2014**, *119*, 383–397. [CrossRef]
15. Zhang, X.M.; Wang, Y.; Lv, X.M.; Li, H.; Sun, P.; Lu, H.; Li, F.L. NtCP56, a new cysteine protease in Nicotiana tabacum L., involved in pollen grain development. *J. Exp. Bot.* **2009**, *60*, 1569–1577. [CrossRef]

16. Lee, S.; Jung, K.H.; An, G.; Chung, Y.Y. Isolation and characterization of a rice cysteine protease gene, OsCP1, using T-DNA gene-trap system. *Plant Mol. Biol.* **2004**, *54*, 755–765. [CrossRef]
17. Li, Y.; Niu, L.; Zhou, X.; Liu, H.; Tai, F.; Wang, W. Modifying the Expression of Cysteine Protease Gene PCP Affects Pollen Development, Germination and Plant Drought Tolerance in Maize. *Int. J. Mol. Sci.* **2023**, *24*, 7406. [CrossRef]
18. Chunxia, G. Function of Rice Cysteine Protease Gene OsCP6 in Seed Germination and Early Seedling Growth. Master's Thesis, Nanjing Agricultural University, Nanjing, China, 2022.
19. Chen, H.J.; Su, C.T.; Lin, C.H.; Huang, G.J.; Lin, Y.H. Expression of sweet potato cysteine protease SPCP2 altered developmental characteristics and stress responses in transgenic Arabidopsis plants. *J. Plant Physiol.* **2010**, *167*, 838–847. [CrossRef]
20. Jahn, R.; Cafiso, D.C.; Tamm, L.K. Mechanisms of SNARE proteins in membrane fusion. *Nat. Rev. Mol. Cell Biol.* **2024**, *25*, 101–118. [CrossRef]
21. Takebe, N.; Nakamura, A.; Watanabe, T.; Miyashita, A.; Satoh, S.; Iwai, H. Cell wall Glycine-rich Protein2 is involved in tapetal differentiation and pollen maturation. *J. Plant Res.* **2020**, *133*, 883–895. [CrossRef]
22. Kim, Y.O.; Kim, J.S.; Kang, H. Cold-inducible zinc finger-containing glycine-rich RNA-binding protein contributes to the enhancement of freezing tolerance in Arabidopsis thaliana. *Plant J.* **2005**, *42*, 890–900. [CrossRef]
23. Lakhssassi, N.; Doblas, V.G.; Rosado, A.; Del Valle, A.E.; Posé, D.; Jimenez, A.J.; Castillo, A.G.; Valpuesta, V.; Borsani, O.; Botella, M.A. The Arabidopsis tetratricopeptide thioredoxin-like gene family is required for osmotic stress tolerance and male sporogenesis. *Plant Physiol.* **2012**, *158*, 1252–1266. [CrossRef] [PubMed]
24. Kim, H.J.; Ok, S.H.; Bahn, S.C.; Jang, J.; Oh, S.A.; Park, S.K.; Twell, D.; Ryu, S.B.; Shin, J.S. Endoplasmic reticulum- and Golgi-localized phospholipase A2 plays critical roles in Arabidopsis pollen development and germination. *Plant Cell* **2011**, *23*, 94–110. [CrossRef] [PubMed]
25. Wu, Q.; Yang, J.; Cheng, N.; Hirschi, K.D.; Park, S. Glutaredoxins in plant development, abiotic stress response, and iron homeostasis: From model organisms to crops. *Environ. Exp. Bot.* **2017**, *139*, 91–98. [CrossRef]
26. Dominguez, P.G.; Carrari, F. ASR1 transcription factor and its role in metabolism. *Plant Signal Behav.* **2015**, *10*, e992751. [CrossRef]
27. Zhang, X.; Zheng, S.; Zhu, H.; Liu, Z.; Peng, X. ER-localized adenine nucleotide transporter ER-ANT1: An integrator of energy and stress signaling in rice. *Plant Mol. Biol.* **2016**, *92*, 701–715. [CrossRef]
28. Schneider, M.; Knuesting, J.; Birkholz, O.; Heinisch, J.J.; Scheibe, R. Cytosolic GAPDH as a redox-dependent regulator of energy metabolism. *BMC Plant Biol.* **2018**, *18*, 184. [CrossRef]
29. Tien, Y.C.; Chuankhayan, P.; Huang, Y.C.; Chen, C.D.; Alikhajeh, J.; Chang, S.L.; Chen, C.J. Crystal structures of rice (*Oryza sativa*) glyceraldehyde-3-phosphate dehydrogenase complexes with NAD and sulfate suggest involvement of Phe37 in NAD binding for catalysis. *Plant Mol. Biol.* **2012**, *80*, 389–403. [CrossRef]
30. Lin, D.; Zhang, L.; Mei, J.; Chen, J.; Piao, Z.; Lee, G.; Dong, Y. Mutation of the rice TCM12 gene encoding 2,3-bisphosphoglycerate-independent phosphoglycerate mutase affects chlorophyll synthesis, photosynthesis and chloroplast development at seedling stage at low temperatures. *Plant Biol.* **2019**, *21*, 585–594. [CrossRef]
31. Tao, X.; Wu, Q.; Fu, X.; Zhu, B.; Chen, F.; Liu, B.; Mao, L.; Luo, Z.; Li, L.; Ying, T. Understanding of exogenous auxin in regulating sucrose metabolism during postharvest tomato fruit ripening. *Postharvest Biol. Technol.* **2022**, *189*, 111913. [CrossRef]
32. Lee, M.H.; Jung, C.; Lee, J.; Kim, S.Y.; Lee, Y.; Hwang, I. An Arabidopsis prenylated Rab acceptor 1 isoform, AtPRA1.B6, displays differential inhibitory effects on anterograde trafficking of proteins at the endoplasmic reticulum. *Plant Physiol.* **2011**, *157*, 645–658. [CrossRef]
33. Sun, X.; Ma, Y.; Yang, C.; Li, J. Rice OVATE family protein 6 regulates leaf angle by modulating secondary cell wall biosynthesis. *Plant Mol. Biol.* **2020**, *104*, 249–261. [CrossRef]
34. Borassi, C.; Sede, A.R.; Mecchia, M.A.; Mangano, S.; Marzol, E.; Denita-Juarez, S.P.; Salgado Salter, J.D.; Velasquez, S.M.; Muschietti, J.P.; Estevez, J.M. Proline-rich extensin-like receptor kinases PERK5 and PERK12 are involved in pollen tube growth. *FEBS Lett.* **2021**, *595*, 2593–2607. [CrossRef] [PubMed]
35. Hou, S.; Shi, J.; Hao, L.; Wang, Z.; Liao, Y.; Gu, H.; Dong, J.; Dresselhaus, T.; Zhong, S.; Qu, L.J. VPS18-regulated vesicle trafficking controls the secretion of pectin and its modifying enzyme during pollen tube growth in Arabidopsis. *Plant Cell* **2021**, *33*, 3042–3056. [CrossRef] [PubMed]
36. Guo, H.; Xiao, C.; Liu, Q.; Li, R.; Yan, Z.; Yao, X.; Hu, H. Two galacturonosyltransferases function in plant growth, stomatal development, and dynamics. *Plant Physiol.* **2021**, *187*, 2820–2836. [CrossRef]
37. Shi, Y.Z.; Zhu, X.F.; Miller, J.G.; Gregson, T.; Zheng, S.J.; Fry, S.C. Distinct catalytic capacities of two aluminium-repressed Arabidopsis thaliana xyloglucan endotransglucosylase/hydrolases, XTH15 and XTH31, heterologously produced in Pichia. *Phytochemistry* **2015**, *112*, 160–169. [CrossRef] [PubMed]
38. Zhu, Q.; Shao, Y.; Ge, S.; Zhang, M.; Zhang, T.; Hu, X.; Liu, Y.; Walker, J.; Zhang, S.; Xu, J. A MAPK cascade downstream of IDA-HAE/HSL2 ligand-receptor pair in lateral root emergence. *Nat. Plants* **2019**, *5*, 414–423. [CrossRef]
39. Lehti-Shiu, M.D.; Shiu, S.H. Diversity, classification and function of the plant protein kinase superfamily. *Philos. Trans. R. Soc. Lond. B Biol. Sci.* **2012**, *367*, 2619–2639. [CrossRef]

 https://doi.org/10.3390/plants15050677

40. Zhu, Q.; Tan, Q.; Gao, Q.; Zheng, S.; Chen, W.; Galaud, J.P.; Li, X.; Zhu, X. Calmodulin-like protein CML15 interacts with PP2C46/65 to regulate papaya fruit ripening via integrating calcium, ABA and ethylene signals. *Plant Biotechnol. J.* **2024**, *22*, 1703–1723. [CrossRef]
41. Schmittgen, T.D. Analyzing Real-time PCR data by the comparative CT method. *Nat. Protoc.* **2008**, *3*, 1101 1108. [CrossRef]
42. Wang, C.; Gao, B.; Chen, N.; Jiao, P.; Jiang, Z.; Zhao, C.; Ma, Y.; Guan, S.; Liu, S. A Novel Senescence-Specific Gene (ZmSAG39) Negatively Regulates Darkness and Drought Responses in Maize. *Int. J. Mol. Sci.* **2022**, *23*, 15984. [CrossRef]
43. Lopez, L.; Camas, A.; Shivaji, R.; Ankala, A.; Williams, P.; Luthe, D. Mir1-CP, a novel defense cysteine protease accumulates in maize vascular tissues in response to herbivory. *Planta* **2007**, *226*, 517–527. [CrossRef] [PubMed]
44. Mueller, A.N.; Ziemann, S.; Treitschke, S.; Aßmann, D.; Doehlemann, G. Compatibility in the Ustilago maydis-maize interaction requires inhibition of host cysteine proteases by the fungal effector Pit2. *PLoS Pathog.* **2013**, *9*, e1003177. [CrossRef] [PubMed]
45. Li, Y.; Niu, L.; Wu, X.; Faleri, C.; Tai, F.; Zhang, M.; Liu, H.; Wang, W.; Cai, G. Genome-Wide Identification and Comparison of Cysteine Proteases in the Pollen Coat and Other Tissues in Maize. *Front. Plant Sci.* **2021**, *12*, 709534. [CrossRef] [PubMed]
46. Si, W. Analysis of Evolution, Localization and Function of Pollen Coat Cysteine Protease of Maize. Ph.D. Dissertation, Henan Agricultural University, Zhengzhou, China, 2018.
47. Rodríguez-García, D.R.; Rondón Guerrero, Y.D.C.; Ferrero, L.; Rossi, A.H.; Miglietta, E.A.; Aptekmann, A.A.; Marzol, E.; Martínez Pacheco, J.; Carignani, M.; Berdion Gabarain, V.; et al. Transcription factor NAC1 activates expression of peptidase-encoding AtCEPs in roots to limit root hair growth. *Plant Physiol.* **2023**, *194*, 81–93. [CrossRef]
48. Luo, J.; Zhang, H.; He, W.; Zhang, Y.; Cao, W.; Zhang, H.; Bao, Y. OsSNAP32, a SNAP25-type SNARE protein-encoding gene from rice, enhanced resistance to blast fungus. *Plant Growth Regul.* **2016**, *80*, 37–45. [CrossRef]
49. Bao, Y.M.; Wang, J.F.; Huang, J.; Zhang, H.S. Molecular cloning and characterization of a novel SNAP25-type protein gene OsSNAP32 in rice (*Oryza sativa* L.). *Mol. Biol. Rep.* **2008**, *35*, 145–152. [CrossRef]
50. Macgregor, S.R.; Beronilla, P.K.S.; Goring, D.R. The Arabidopsis SNARE complex genes regulate the early stages of pollen-stigma interactions. *Plant Reprod* **2024**, *37*, 309–320. [CrossRef]
51. Goring, D.R. Exocyst, exosomes, and autophagy in the regulation of Brassicaceae pollen-stigma interactions. *J. Exp. Bot.* **2017**, *69*, 69–78. [CrossRef]
52. Li, Y.; Liu, P.; Mei, L.; Jiang, G.; Lv, Q.; Zhai, W.; Li, C. Knockout of a papain-like cysteine protease gene OCP enhances blast resistance in rice. *Front. Plant Sci.* **2022**, *13*, 1065253. [CrossRef]
53. Liu, L.; Jia, L.; Xu, L.; Deng, L.; Zhao, P.; Zeng, W. Mutation of iPGAM using the CRISPR/Cas9 system affects pollen vitality and chlorophyll synthesis in Nicotiana tabacum. *Plant Biotechnol. Rep.* **2021**, *15*, 217–227. [CrossRef]
54. Cao, Y.; Yang, W.; Ma, J.; Cheng, Z.; Zhang, X.; Liu, X.; Wu, X.; Zhang, J. An Integrated Framework for Drought Stress in Plants. *Int. J. Mol. Sci.* **2024**, *25*, 9347. [CrossRef]
55. Molendijk, A.J.; Ruperti, B.; Singh, M.K.; Dovzhenko, A.; Ditengou, F.A.; Milia, M.; Westphal, L.; Rosahl, S.; Soellick, T.R.; Uhrig, J.; et al. A cysteine-rich receptor-like kinase NCRK and a pathogen-induced protein kinase RBK1 are Rop GTPase interactors. *Plant J.* **2008**, *53*, 909–923. [CrossRef] [PubMed]
56. Zhang, H.; Zhai, N.; Ma, X.; Zhou, H.; Cui, Y.; Wang, C.; Xu, G. Overexpression of OsRLCK241 confers enhanced salt and drought tolerance in transgenic rice (*Oryza sativa* L.). *Gene* **2021**, *768*, 145278. [CrossRef] [PubMed]
57. Kaur, V.; Yadav, S.K.; Wankhede, D.P.; Pulivendula, P.; Kumar, A.; Chinnusamy, V. Cloning and characterization of a gene encoding MIZ1, a domain of unknown function protein and its role in salt and drought stress in rice. *Protoplasma* **2020**, *257*, 475–487. [CrossRef] [PubMed]
58. Miyazawa, Y.; Moriwaki, T.; Uchida, M.; Kobayashi, A.; Fujii, N.; Takahashi, H. Overexpression of MIZU-KUSSEI1 enhances the root hydrotropic response by retaining cell viability under hydrostimulated conditions in Arabidopsis thaliana. *Plant Cell Physiol.* **2012**, *53*, 1926–1933. [CrossRef]
59. Diaz-Mendoza, M.; Dominguez-Figueroa, J.D.; Velasco-Arroyo, B.; Cambra, I.; Gonzalez-Melendi, P.; Lopez-Gonzalvez, A.; Garcia, A.; Hensel, G.; Kumlehn, J.; Diaz, I.; et al. HvPap-1 C1A Protease and HvCPI-2 Cystatin Contribute to Barley Grain Filling and Germination. *Plant Physiol.* **2016**, *170*, 2511–2524. [CrossRef]
60. Liu, J.; Hasanuzzaman, M.; Wen, H.; Zhang, J.; Peng, T.; Sun, H.; Zhao, Q. High temperature and drought stress cause abscisic acid and reactive oxygen species accumulation and suppress seed germination growth in rice. *Protoplasma* **2019**, *256*, 1217–1227. [CrossRef]
61. Cercós, M.; Gómez-Cadenas, A.; Ho, T.H. Hormonal regulation of a cysteine proteinase gene, EPB-1, in barley aleurone layers: Cis- and trans-acting elements involved in the co-ordinated gene expression regulated by gibberellins and abscisic acid. *Plant J.* **1999**, *19*, 107–118. [CrossRef]
62. Yang, X.; Feng, J.; Xu, K.; Han, J.; Zhang, M.; Cheng, A.; Wang, Z.; Wang, F.; Zhou, X.; Jiang, L.; et al. RD21 enhances resistance to the strawberry vein banding virus by promoting autophagy-mediated degradation of the viral silencing suppressor P6 protein. *Plant Biotechnol. J.* **2025**, *23*, 3597–3611. [CrossRef]

 https://doi.org/10.3390/plants15050677

63. Chen, C.; Chen, H.; Zhang, Y.; Thomas, H.R.; Frank, M.H.; He, Y.; Xia, R. TBtools: An Integrative Toolkit Developed for Interactive Analyses of Big Biological Data. *Mol. Plant.* **2020**, *13*, 1194–1202. [CrossRef]
64. Livak, K.J.; Schmittgen, T.D. Analysis of Relative Gene Expression Data Using Real-Time Quantitative PCR and the the $2^{-\Delta\Delta CT}$ Method. *Methods* **2001**, *25*, 402–408. [CrossRef] [PubMed]
65. Rolland, V. Determining the Subcellular Localization of Fluorescently Tagged Proteins Using Protoplasts Extracted from Transiently Transformed Nicotiana benthamiana Leaves. *Methods Mol. Biol.* **2018**, *1770*, 263–283. [CrossRef] [PubMed]
66. Liang, B.; Cao, J.; Wang, R.; Fan, C.; Wang, W.; Hu, X.; He, R.; Tai, F. ZmCIPK32 positively regulates germination of stressed seeds via gibberellin signal. *Plant Physiol. Biochem.* **2023**, *199*, 107716. [CrossRef] [PubMed]
67. Zhao, Y.; Du, H.; Wang, Y.; Wang, H.; Yang, S.; Li, C.; Chen, N.; Yang, H.; Zhang, Y.; Zhu, Y.; et al. The calcium-dependent protein kinase ZmCDPK7 functions in heat-stress tolerance in maize. *J. Integr. Plant Biol.* **2021**, *63*, 510–527. [CrossRef]
68. Du, G.; Xu, J.; Gao, C.; Lu, J.; Li, Q.; Du, J.; Lv, M.; Sun, X. Effect of low storage temperature on pollen viability of fifteen herbaceous peonies. *Biotechnol. Rep.* **2019**, *21*, e00309. [CrossRef]
69. Zhu, Y.; Zhao, P.; Wu, X.; Wang, W.; Scali, M.; Cresti, M. Proteomic identification of differentially expressed proteins in mature and germinated maize pollen. *Acta Physiol. Plantarum.* **2011**, *33*, 1467–1474. [CrossRef]
70. Lausser, A.; Kliwer, I.; Srilunchang, K.O.; Dresselhaus, T. Sporophytic control of pollen tube growth and guidance in maize. *J. Exp. Bot.* **2010**, *61*, 673–682. [CrossRef]
71. Ferjani, A.; Tsukagoshi, H.; Vassileva, V. Editorial: Model organisms in plant science: Arabidopsis thaliana. *Front. Plant Sci.* **2023**, *14*, 1279230. [CrossRef]

 https://doi.org/10.3390/plants15050677

plants

MDPI

Article

CEHD: A Unified Framework for Detection and Height Estimation of Fresh Corn Ears in Field Conditions

Hengyi Wang [1,2,3], Yang Li [1,2,3], Jun Fu [1,2,3], Qiankun Fu [1,2,*] and Yongliang Qiao [4]

1 College of Biological and Agricultural Engineering, Jilin University, Changchun 130022, China; hengyi23@mails.jlu.edu.cn (H.W.); liyang20210706@163.com (Y.L.); fu_jun@jlu.edu.cn (J.F.)
2 Key Laboratory of Efficient Sowing and Harvesting Equipment, Ministry of Agriculture and Rural Affairs, Jilin University, Changchun 130022, China
3 Key Laboratory of Bionic Engineering, Ministry of Education, Jilin University, Changchun 130022, China
4 Australian Institute Machine Learning (AIML), University of Adelaide, Adelaide, SA 5005, Australia; yongliang.qiao@ieee.org
* Correspondence: qkfu@jlu.edu.cn

Abstract

Real-time detection of fresh corn ear height can provide a basis for dynamic adjustment of harvester header parameters, reducing mechanical damage and improving harvest quality. This study proposes a corn ear height detection model (CEHD). A YOLO-HAMDF network is developed for ear recognition, in which the core modules—TBDA, GLSA, and AQE—respectively suppress background interference, enhance contextual perception, and optimize bounding-box scoring. Depth information is incorporated to filter non-target regions and improve system robustness. In addition, a DI-DeepSORT module is designed for ear tracking, where DBC-Net and IDA-Kalman, respectively, enhance the discriminability of ReID features and enable independent-dimension adaptive noise modeling with smoothed positional updates. Experimental results demonstrate that the proposed CEHD model achieves a mean absolute error (MAE) of only 3.21 ± 0.05 cm under field conditions, indicating strong stability and practical applicability. In summary, this study presents a stable and reliable corn ear height detection system, achieves real-time monitoring of ear height, and provides data support for the dynamic adjustment of header parameters in fresh corn harvesters.

Keywords: fresh corn; YOLO-HAMDF; DeepSORT; RGB-D object detection; height measurement

1. Introduction

Fresh corn is a high-value crop used for both vegetable and grain purposes and holds significant economic importance [1]. However, its kernels typically contain more than 60% moisture at harvest [2]. Their soft texture and low resistance to mechanical impact make them highly susceptible to breakage during mechanized harvesting [3,4], which severely reduces product quality and economic return. With the advancement of agricultural intelligence and digitalization—particularly within the Agriculture 5.0 framework—the real-time and accurate acquisition of crop phenotypic traits, such as ear characteristics, has become an effective means to guide machinery parameter adjustment and reduce harvest-induced damage [5]. Among these traits, ear height is a key phenotypic indicator. Its accurate measurement directly supports the dynamic adjustment of harvester header parameters, thereby minimizing ear damage under complex field conditions [6].

As the primary prerequisite for height measurement, object detection has attracted extensive research interest worldwide. Compared with traditional detection methods such as color-difference analysis [7], K-means clustering [8], fuzzy C-means [9], KNN-based approaches [10], and SVM-based classification [11], deep learning–based object detection techniques offer stronger feature extraction capabilities and markedly improved real-time performance. Current detection frameworks are primarily based on Transformer architectures and convolutional neural networks (CNNs). Representative Transformer-based approaches include Vision Transformer [12], Swin Transformer [13], and RT-DETR [14]. Wu et al. developed CBAM-RT-DETR, a real-time detector that integrates CBAM and grouped convolutions to enhance shallow features, improving maize seedling detection in complex field environments [15]. Simşek Bagcı et al. proposed a crop-recognition method based on a Deep Transformer Encoder that uses multi-temporal Sentinel-1 and Landsat-8 imagery to automatically identify cotton and corn planting areas, achieving 85–95% classification accuracy across different temporal combinations [16]. Zhang et al. introduced an MConv-SwinT–based maize quality detection model that fuses shallow and deep features and incorporates a convolutional block attention mechanism. The model achieved 99.89% accuracy in classifying maize quality levels [17]. Although Transformer-based models effectively integrate multi-scale information and handle occlusions, their strong emphasis on long-range dependencies often results in the loss of shallow detail information. In addition, their high computational cost and large parameter size pose major challenges for embedded deployment and real-time applications [18]. In contrast, lightweight CNN-based detectors provide efficient and stable performance in resource-constrained environments while maintaining high computational efficiency [19]. Among CNN-based detectors, the YOLO family has become highly influential due to its strong real-time performance, high accuracy, and compact architecture. For instance, Gai et al. integrated DenseNet into YOLOv4 and designed circular annotation boxes to detect cherries at different maturity stages with high precision [20]. Chen et al. proposed GE-YOLO based on YOLOv8, leveraging a Gold-YOLO structure and EMA attention to enhance the robustness of rice-weed recognition [21]. Zhou et al. introduced YOLO-ACE, which incorporates a contextual enhancement module and selective convolutional attention to improve cotton-field weed detection [22]. Sun et al. proposed ESC-YOLO by integrating EfficientViT with SCConv, improving apple fruit recognition performance [23]. Considering the requirements for real-time performance and embedded deployment, this study adopts the lightweight and mature YOLO11-L as the baseline model for corn ear detection.

However, two-dimensional object detection provides only planar information and cannot obtain the height measurements required for adjusting harvester header parameters. Depth cameras, as common sensors for acquiring three-dimensional visual information, are insensitive to lighting conditions, robust to complex backgrounds, and capable of providing highly stable depth data [24]. When integrated with object detection, they allow reliable estimation of the three-dimensional position of corn ears, thereby enabling accurate ear-height measurement. Zhong et al. combined Faster R-CNN with depth information to predict pepper plant height, achieving an error within 4.4 mm [25]. Baden Parr et al. utilized RGB and ToF depth cameras together with YOLO7 to detect grape berries and estimate their three-dimensional characteristics, obtaining a counting correlation coefficient of 0.946 [26,27]. Zhao et al. integrated an improved YOLO8n-seg model with depth information to measure lettuce height, achieving accuracies of 94.339% in hydroponic systems and 91.22% in potted scenarios [28]. These studies demonstrate notable progress in crop-height measurement.

Most height-measurement studies still focus on improving object detection, with insufficient attention given to object tracking. The stability of the detected bounding boxes

 https://doi.org/10.3390/plants15010038

directly affects the accuracy of depth information extraction [29]. Introducing tracking algorithms can therefore further improve the accuracy and stability of height estimation. Existing studies have shown that integrating object detection with tracking can yield strong performance: Wang et al. integrated YOLO with SORT for mango tracking and counting [30]; Li et al. combined YOLO8 with ByteTrack to build a real-time maize kernel-shedding counting system [31]; Ye et al. integrated an improved YOLO5 with StrongSORT, employing attention mechanisms and weighted box fusion to achieve high-precision tracking and counting of densely infected trees [32]; Tu et al. applied lightweight YOLO5s with an improved DeepSORT, utilizing trajectory delay and secondary IoU matching to enable robust fruit detection, tracking, and counting in complex environments [33]. However, different tracking algorithms exhibit notable limitations. SORT relies solely on motion cues, making it prone to ID switching in occlusion or dense-target scenarios. StrongSORT increases computational complexity and is less suitable for embedded deployment. ByteTrack improves recall for missing detections but struggles to distinguish small, visually similar targets [34]. In contrast, DeepSORT integrates motion and appearance cues, providing a favorable balance between ID consistency and computational efficiency, and thus serves as a strong baseline for further improvement. Currently, tracking-algorithm enhancements specifically targeting crop height measurement remain limited. Most existing studies employ general-purpose tracking models, which do not fully complement the enhanced detection modules and thus fail to form a specialized height-measurement system tailored to the characteristics of specific crops and field environments.

Complex backgrounds and color-similar interference can significantly reduce object detection accuracy in fresh corn harvesting scenarios. Occlusion and the close spatial arrangement of ears often blur ear boundaries and cause false detections. Ears located farther from the camera appear smaller and are more susceptible to missed detections and spurious high-confidence boxes. Harvester-induced motion introduces positional fluctuations across consecutive frames, leading to discontinuous detection outputs and reduced height-measurement accuracy. To address these challenges, a dedicated corn ear dataset was constructed. Unlike public datasets, it simulates real harvesting conditions in both camera angle and installation height, and simultaneously records depth information aligned with RGB images. Based on this dataset, a corn ear height detection framework integrating YOLO-HAMDF and DI-DeepSORT was developed. High-precision ear localization under complex field conditions is achieved through three modules: the TBDA module, which incorporates multi-scale dynamic convolution and triple-attention fusion; the GLSA module, which enables collaborative modeling of local and global features; and the LQE, which integrates occlusion cues and dynamic score suppression. In addition, the DBC-Net architecture and the proposed IDA-Kalman algorithm enhance the DeepSORT framework, improving tracking stability and accuracy. Combined with depth information, the overall framework provides robust ear detection and reliable height estimation in challenging harvesting environments. The main contributions of this study are as follows: (1) A YOLO-HAMDF model was designed, significantly improving corn ear recognition accuracy under complex field conditions; (2) A DI-DeepSORT model was introduced to address the limited feature-extraction capacity and non-adaptive observation-noise modeling of traditional DeepSORT, thereby improving cross-frame tracking consistency; (3) A corn ear height detection model (CEHD) was developed by integrating depth information, enabling real-time monitoring of ear height in the field.

https://doi.org/10.3390/plants15010038

2. Experimental Results and Analysis

2.1. Ablation Study

To evaluate the contribution of each module to corn ear height detection, ablation experiments were conducted on the self-constructed fresh corn ear dataset (FCE-GBD Dataset). YOLO11-L combined with the traditional DeepSORT was used as the baseline, and each module as well as their combinations were tested multiple times. Five random seeds were applied for each experiment to assess result stability, with both mean values and standard deviations reported. The results are summarized in Table 1.

Table 1. Ablation studies on the FCE-GBD Dataset.

Baseline	TBDA [1]	GLSA [2]	AQE [3]	DI- Deep SORT [4]	mAP0.5 (%)	MAE (cm)	Pearson r	FLOPs (G)	Latency (ms)	FPS
√	×	×	×	×	93.21 ± 0.18	6.52 ± 0.12	0.83 ± 0.01	**86.9**	**32.1 ± 0.4**	**31 ± 0.4**
√	√	×	×	×	96.35 ± 0.12	5.26 ± 0.08	0.86 ± 0.01	87.4	33.5 ± 0.5	30 ± 0.5
√	×	√	×	×	96.19 ± 0.15	5.19 ± 0.10	0.85 ± 0.01	87.3	33.2 ± 0.5	30 ± 0.5
√	×	×	√	×	95.21 ± 0.10	4.15 ± 0.07	0.90 ± 0.01	87.1	32.8 ± 0.4	30 ± 0.4
√	×	×	×	√	93.21 ± 0.18	5.56 ± 0.09	0.84 ± 0.01	87.2	33.0 ± 0.4	30 ± 0.4
√	√	√	×	×	96.80 ± 0.10	4.90 ± 0.07	0.87 ± 0.01	87.8	34.6 ± 0.4	29 ± 0.4
√	×	√	√	×	96.75 ± 0.10	4.05 ± 0.06	0.91 ± 0.01	87.5	33.9 ± 0.4	29 ± 0.4
√	√	×	√	×	96.90 ± 0.09	4.00 ± 0.05	0.91 ± 0.01	87.6	34.2 ± 0.5	29 ± 0.4
√	√	√	√	×	**97.01 ± 0.08**	3.50 ± 0.05	0.92 ± 0.01	88.0	35.3 ± 0.5	28 ± 0.4
√	√	√	√	√	**97.01 ± 0.08**	**3.21 ± 0.05**	**0.94 ± 0.01**	88.3	36.5 ± 0.5	28 ± 0.4

[1] Three-Branch Dynamic Attention module. [2] Global-Local Selective Aggregation module. [3] Adaptive Quality Estimator. [4] DBC-IDA DeepSORT. "√" = module present, "×" = module absent.

For detection accuracy (mAP0.5), introducing individual modules TBDA, GLSA, and AQE increased mAP0.5 to 96.35 ± 0.12%, 96.19 ± 0.15%, and 95.21 ± 0.10%, respectively, with TBDA yielding the largest improvement. For pairwise combinations, mAP0.5 reached 96.80 ± 0.10% for TBDA + GLSA, 96.75 ± 0.10% for TBDA + AQE, and 96.90 ± 0.09% for GLSA + AQE, demonstrating the synergistic effect of modules. The three-module combination (TBDA + GLSA + AQE) achieved the highest mAP0.5 of 97.01 ± 0.08%, which remained unchanged when all modules were employed simultaneously.

In terms of detection error (MAE) and correlation (Pearson r), single-module MAEs were 5.26 ± 0.08 cm (TBDA), 5.19 ± 0.10 cm (GLSA), 4.15 ± 0.07 cm (AQE), and 5.56 ± 0.09 cm (DI-DeepSORT). Pairwise combinations yielded MAEs of 4.90 ± 0.07 cm (TBDA + GLSA), 4.05 ± 0.06 cm (TBDA + AQE), and 4.00 ± 0.05 cm (GLSA + AQE), with Pearson r ranging from 0.87 to 0.91 ± 0.01. The three-module combination achieved an MAE of 3.50 ± 0.05 cm and Pearson r of 0.92 ± 0.01, while the full-module combination yielded the lowest MAE of 3.21 ± 0.05 cm and the highest Pearson r of 0.94 ± 0.01.

Regarding inference efficiency, introducing single modules had minimal impact on FLOPs (87.1–87.4 G), inference latency (32.8–33.5 ms), and frame rate (30–31 FPS). When all modules were integrated, FLOPs increased to 88.3 G, latency rose to 36.5 ms, and frame rate decreased slightly to 28 FPS. These results demonstrate a balance between high accuracy and real-time performance suitable for field deployment.

Specifically, the three improved modules in YOLO-HAMDF each serve distinct functions. The TBDA module employs multi-scale dynamic convolution combined with a triple attention mechanism—SE channel attention, spatial attention, and energy function attention—to fuse features across different receptive fields. This design enables the model to capture both fine-grained variations and spatial structural characteristics of corn ears while emphasizing foreground information and suppressing background interference from leaves and weeds. As a result, adaptability and bounding-box stability are significantly enhanced under complex scenarios. The GLSA module splits input features into local and global branches. The local branch leverages depthwise separable convolutions, residual

 https://doi.org/10.3390/plants15010038

connections, and attention mechanisms to strengthen detailed representation in key regions of the corn ears. The global branch captures long-range spatial dependencies and supplements global semantic information. Together, these branches optimize feature representation and contextual awareness. The AQE introduces an occlusion factor, dynamic output, and a staged score fusion strategy. It effectively suppresses falsely high-confidence boxes for occluded or distant targets, reducing misdetections and false positives. The staged fusion strategy applies additive fusion during early training and transitions to multiplicative fusion in later stages, improving training convergence and enhancing suppression of low-quality targets. Collectively, the three modules enable the model to achieve higher robustness and precision in dense, occluded, and scale-diverse field environments. They provide a stable and reliable foundation for corn ear localization and height measurement.

Furthermore, the DI-DeepSORT plays a critical role in corn ear height detection. For feature extraction, DBC-Net replaces the original CNN and employs multi-scale convolution combined with channel–spatial attention to effectively highlight salient corn ear features, enhancing cross-frame matching robustness. In addition, the traditional Kalman filter is enhanced by breaking the constraint of fixed observation noise modeling. IDA-Kalman dynamically adjusts the observation noise covariance matrix based on detection confidence and per-dimension weighting, improving adaptability to occlusions and low-confidence observations. Moreover, the introduced exponential smoothing mechanism reduces jitter in the center position, further enhancing tracking stability and continuity.

2.2. Comparison Experiments

2.2.1. Comparison of Detection Performances

To validate the effectiveness of the proposed method for fresh corn ear detection, a comparative evaluation was conducted on the FCE-GBD Dataset. The improved model was evaluated against 17 mainstream object detection algorithms. These included CNN-based detectors such as Faster R-CNN [35], Cascade R-CNN [36], YOLO5-L [37], YOLO8-L, YOLO11-L [38], and YOLO13-L [39], as well as Transformer-based detectors, including RT-DETRv3 [14] and Swin-T [13]. The results of the comparison are summarized in Table 2.

Table 2. Comparison of Detection Performances.

Method	mAP@[0.5:0.95] (%)	mAP (%)	mAP@0.75 (%)	AP_S (%)	AP_M (%)	AP_L (%)	Parameter (M)	FLOPs (G)
Faster R-CNN	74.32	79.24	68.50	52.1	76.3	88.5	41.12	216.30
Cascade R-CNN	75.20	79.50	69.20	53.8	76.9	88.5	68.93	244.10
YOLO5-L	85.12	89.47	80.33	68.5	84.7	92.1	46.11	107.6
YOLO8-L	88.50	92.35	84.20	72.0	87.5	94.0	43.61	164.82
RT-DETRv3	87.20	91.50	83.10	70.5	86.0	93.0	52.0	180.0
Swin-T	88.10	92.70	83.90	65.0	87.0	94.5	60.0	210.0
YOLO11-L	89.21	93.21	85.50	73.2	88.2	94.5	**25.3**	**86.9**
YOLO13-L	90.05	94.10	86.70	74.0	89.0	95.0	28.0	92.0
YOLO-HAMDF	**93.50**	**97.01**	**91.20**	**82.1**	**93.0**	**97.5**	26.4	89.1

The experimental results indicate that among CNN-based detection methods, Faster R-CNN and Cascade R-CNN achieved mAP0.5 values of 79.24% and 79.50%, respectively. These models demonstrated relatively low accuracy and high computational complexity (216.3–244.1 G FLOPs). The YOLO series performed better. YOLO5-L, YOLO8-L, YOLO11-L, and YOLO13-L achieved mAP0.5 values of 89.47%, 92.35%, 93.21%, and 94.10%, respectively, while maintaining a modest parameter count of 25.3–46.1 M and FLOPs ranging from 86.9 to 164.8 G. This demonstrates a balance between accuracy and inference efficiency. Transformer-based methods, RT-DETRv3 and Swin-T, achieved slightly higher

 https://doi.org/10.3390/plants15010038

mAP0.5 values of 91.50% and 92.70% compared with YOLO8-L. However, their larger parameter sizes (52–60 M) and higher computational costs (180–210 G FLOPs) resulted in relatively lower efficiency. The proposed YOLO-HAMDF model introduces TBDA and GLSA modules, as well as the AQE, achieving significant performance improvements. It attained mAP0.5 of 97.01%, mAP [0.5:0.95] of 93.50%, and mAP0.75 of 91.20%, with APS, APM, and APL values of 82.1%, 93.0%, and 97.5% for small, medium, and large targets, respectively. The model contains only 26.4 M parameters and 89.1 G FLOPs, outperforming other comparative models in both accuracy and efficiency, fully demonstrating the advantage of multi-module collaborative enhancement.

Figure 1 presents a comparison of confusion matrices across all detection models. Among CNN-based models, Faster R-CNN and Cascade R-CNN achieved recall rates of 0.78 and 0.79, indicating limited capability in detecting small, occluded, or sparsely distributed corn ears in complex field environments. YOLO series models performed better. YOLO5-L, YOLO8-L, YOLO11-L, and YOLO13-L achieved recall rates of 0.87, 0.91, 0.92, and 0.93, demonstrating stronger detection ability for multi-scale targets in complex backgrounds. Transformer-based RT-DETRv3 and Swin-T achieved recall rates of 0.90 and 0.91, showing good performance for large targets but slightly weaker recognition for small or sparse corn ears. The proposed YOLO-HAMDF achieved the highest recall rate of 0.95, representing an improvements of approximately 4% and 2% compared with Swin-T and YOLO13-L, respectively. The combination of high recall and low miss rate highlights the robustness and reliability of YOLO-HAMDF for corn ear detection under complex field conditions.

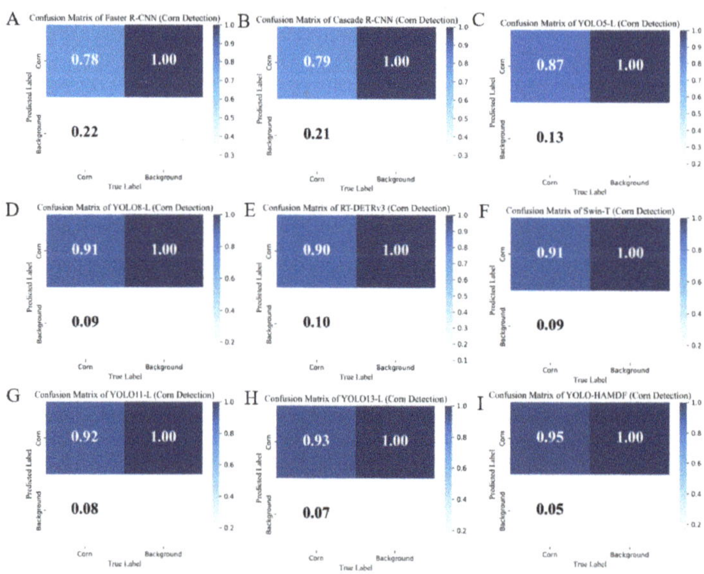

Figure 1. Comparison of confusion matrices of different models evaluated on FCE-GBD. (**A**) Faster R-CNN. (**B**) Cascade R-CNN. (**C**) YOLO5-L. (**D**) YOLO8-L. (**E**) RT-DETRv3. (**F**) Swin-T. (**G**) YOLO11-L. (**H**) YOLO13-L. (**I**) YOLO-HAMDF.

2.2.2. Comparison of Tracking Performances

To objectively evaluate the performance of the improved IDA Kalman filter in corn ear tracking, the tracking results of three methods were compared under identical detection conditions: Raw Detection (direct outputs of the YOLO11 detector), Original Kalman (the conventional Kalman filter), and IDA Kalman (the improved adaptive filter). All tests em-

https://doi.org/10.3390/plants15010038

ployed the YOLO-HAMDF model for corn ear detection. As shown in the Euclidean error curves (Figure 2), the Raw Detection method exhibited the largest positional errors, with obvious jitter and noise. The Original Kalman filter partially suppressed detection noise, reducing errors; however, significant fluctuations were still observed during occlusion or low-confidence observation periods (frames 40–60). In contrast, the IDA Kalman filter substantially reduced overall errors and effectively suppressed fluctuations. Quantitative analysis indicates that during the occlusion period, the root mean square error (RMSE) of IDA Kalman was only 3.5 pixels, compared with 8.2 pixels for the Original Kalman filter and 12.0 pixels for Raw Detection.

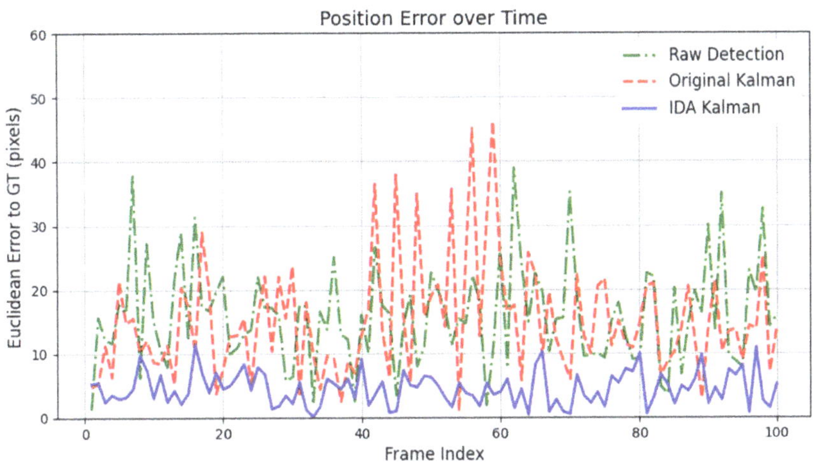

Figure 2. Euclidean position error over frames for corn ear tracking with different filtering strategies.

Furthermore, the proposed method was compared with DeepSORT, ByteTrack, and StrongSORT. To ensure a fair comparison, all four tracking methods were evaluated using the detection outputs from the YOLO11-L detector. As shown in Table 3, the proposed method achieved a MOTA of 92%, outperforming DeepSORT (88%), ByteTrack (86%), and StrongSORT (89%). This indicates that the introduction of the IDA Kalman filter significantly improved the accuracy of target state prediction and association. Regarding the IDF1 metric, the proposed method achieved 90%, higher than DeepSORT (85%), ByteTrack (82%), and StrongSORT (87%). This demonstrates enhanced identity preservation and effectively reduces target drift. The number of ID switches was only 15, considerably lower than DeepSORT (30), ByteTrack (40), and StrongSORT (20), further confirming stable tracking performance under occlusion and dense-target scenarios. Moreover, the proposed method maintained high tracking accuracy while achieving a real-time processing speed of 30 FPS, indicating a well-balanced trade-off between efficiency and performance.

Table 3. Comparison of Tracking Performances.

Tracker	MOTA (%)	IDF1 (%)	ID Switch	FPS
DI- DeepSORT	**92**	**90**	**15**	30
DeepSORT	88	85	30	**31**
ByteTrack	86	82	40	22
StrongSORT	89	87	20	15

Overall, by integrating motion information, appearance features, and independent-dimension adaptive noise modeling, the proposed method effectively improved corn ear tracking accuracy in complex field environments. Moreover, it enhanced the stability

https://doi.org/10.3390/plants15010038

of height estimation, demonstrating a clear performance advantage over conventional DeepSORT, ByteTrack, and StrongSORT.

2.2.3. Comparison of Height Measurement Performances

To comprehensively assess the performance of the corn ear height measurement system, multiple detector–tracker combinations were evaluated. Three object detectors—YOLO-HAMDF, YOLO13, and Swin-T—were selected, together with two trackers, DBC-IDA-Deep-SORT and StrongSORT. The comparative results of these combinations are presented in Table 4. The selection of YOLO13, Swin-T, and StrongSORT was based on the preceding comparative experiments (Sections 2.2.1 and 2.2.2). Specifically, YOLO13 achieved the best performance among CNN-based detectors, Swin-T performed best among Transformer-based detectors, and StrongSORT demonstrated the strongest tracking capability.

Table 4. Comparison of Height Measurement Performances.

Detectors	Trackers	MAE (cm)	Pearson r	Latency (ms)	FPS
Swin-T	StrongSORT	3.98 ± 0.07	0.875 ± 0.02	53.0 ± 0.8	18.9 ± 0.3
Swin-T	DI-DeepSORT	3.91 ± 0.07	0.88 ± 0.02	49.9 ± 0.7	20.0 ± 0.3
YOLO13	StrongSORT	3.60 ± 0.06	0.895 ± 0.01	46.0 ± 0.7	21.7 ± 0.3
YOLO13	DI-DeepSORT	3.51 ± 0.06	0.90 ± 0.01	42.9 ± 0.6	23.3 ± 0.3
YOLO-HAMDF	StrongSORT	3.28 ± 0.05	0.915 ± 0.01	38.5 ± 0.6	26.0 ± 0.4
YOLO-HAMDF	DI-DeepSORT	**3.21 ± 0.05**	**0.92 ± 0.01**	**35.4 ± 0.5**	**28 ± 0.4**

Under the unified use of the CHSM-PCM height estimation method, the performance of each detector–tracker combination is summarized in Table 4. Overall, the detector's feature representation ability and the tracker's appearance discrimination jointly determine both the accuracy and real-time performance of height measurement. Among all combinations, YOLO-HAMDF + DBC-IDA-Deep-SORT achieved the best results (MAE 3.21 ± 0.05 cm, Pearson r 0.92 ± 0.01, latency 35.4 ± 0.5 ms, FPS 28 ± 0.4). This superiority mainly stems from the higher localization accuracy of YOLO-HAMDF in complex field backgrounds and the more stable appearance matching provided by DBC-IDA-Deep-SORT, which together maintain better temporal consistency in the height estimation process. When YOLO-HAMDF is combined with StrongSORT, although the re-identification capability is improved, the significantly higher computational cost leads to increased latency and reduced FPS, resulting in a slight rise in MAE. For combinations using YOLO13, the weaker feature extraction limits detection of occluded or distant corn ears, leading to higher height estimation errors (MAE 3.51–3.60 cm). Although Swin-T possesses a strong global attention mechanism, its performance on small or densely distributed corn ears is inferior. Furthermore, its slower inference speed further degrades overall results, yielding the lowest performance among all combinations (MAE 3.91–3.98 cm, FPS < 20).

In summary, the detector primarily determines height-measurement accuracy, while the tracker mainly affects temporal stability and real-time performance. The combination of YOLO-HAMDF and DBC-IDA-Deep-SORT achieves the optimal balance among accuracy, robustness, and speed, making it the most practical and application-ready solution in this study.

2.3. Visualization

To further validate the effectiveness of the proposed enhancement modules and the overall framework under complex field conditions, a systematic visualization analysis was conducted from multiple perspectives, including feature attention, detection performance, feature distribution, and spatial filtering.

 https://doi.org/10.3390/plants15010038

First, Eigen Grad-CAM was employed to visualize the feature responses of TBDA, GLSA, and AQE within YOLO11-L (Figure 3). The results indicate that, compared with the baseline YOLO11-L model, the TBDA module substantially reduces attention to leaf regions while markedly enhancing attention to mid- and long range corn ears. This demonstrates that TBDA effectively suppresses background interference and strengthens target-focused representation. Although the GLSA and the AQE module exhibit slightly weaker suppression of leaf-related responses than TBDA, their heatmaps still show clear improvements over the baseline. Moreover, all three modules display concentrated and continuous attention distributions over target regions, indicating significantly enhanced feature representation capability. These findings collectively support the improved robustness of the model in accurately detecting and localizing corn ears within complex agricultural environments.

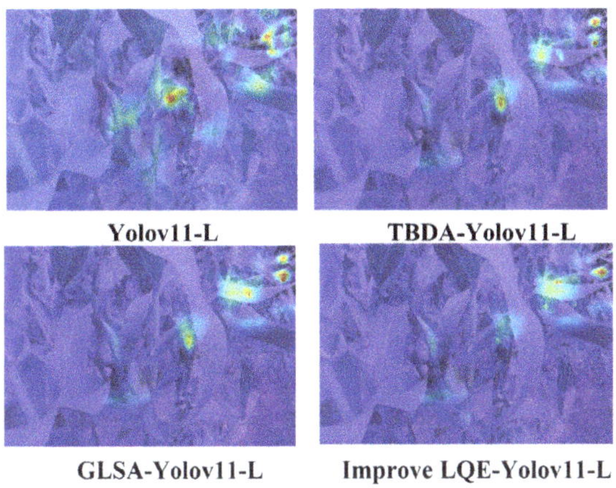

Figure 3. Eigen Grad-CAM visual comparison of feature maps.

Based on the module-enhanced model, the performance of various mainstream detection methods was compared on field images of corn ears (Figure 4). Faster R-CNN and Cascade R-CNN exhibited false positives, missed detections, and duplicate detections under conditions of multiple occluded ears and complex backgrounds. YOLO5-L reduced false and missed detections but still produced duplicate detections when ears were occluded. YOLO8-L further mitigated duplicate detections, though false positives and missed detections remained. Transformer-based detectors, including RT-DETRv3 and Swin-T, effectively handled large ears and suppressed background noise, accurately localizing them. However, both models showed missed detections for small or densely occluded ears. Swin-T slightly outperformed RT-DETRv3 in small-ear localization and recognition but remained inferior to the later YOLO models. YOLO11-L achieved notable overall improvements, largely eliminating false and duplicate detections, though missed detections persisted in scenarios with overlapping or small ears. YOLO13-L further enhanced feature extraction over YOLO11-L, nearly eliminating missed detections, with only minor deviations in ear boundary localization. In contrast, the proposed YOLO-HAMDF model demonstrated superior performance in recognizing distant small ears and heavily occluded ears, effectively reducing missed detections while improving localization accuracy.

The spatial filtering process is visualized in Figure 5. In the original image (Figure 5a), red bounding boxes indicate corn ears in non-target picking rows, while blue boxes denote ears within the target row. Using the depth map acquired by the RGB-D camera (Figure 5b), the spatial distance of each pixel from the camera was extracted. Field measurements show that the row spacing of corn is approximately 0.6 m. Considering that the mobile

 https://doi.org/10.3390/plants15010038

platform travels along the centerline between rows, a depth threshold of 0.5 m was applied to distinguish target rows from non-target rows. This threshold not only fully includes the plants in the target row but also effectively removes plants and ears in non-target rows beyond 0.5 m. Consequently, irrelevant information is eliminated from the original RGB image, resulting in a background-removed image that retains only the ears in the target row (Figure 5c).

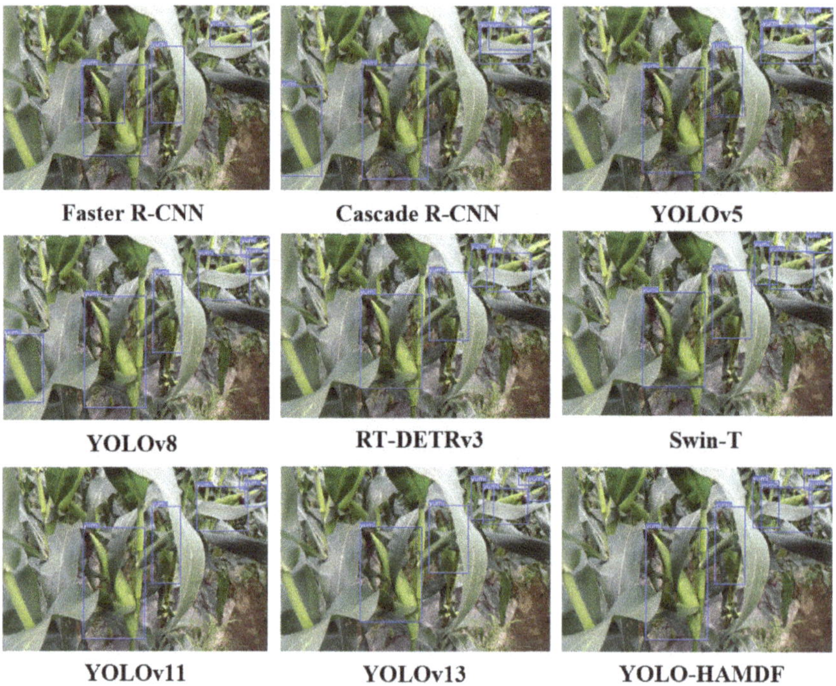

Figure 4. Detection Results of Different Methods.

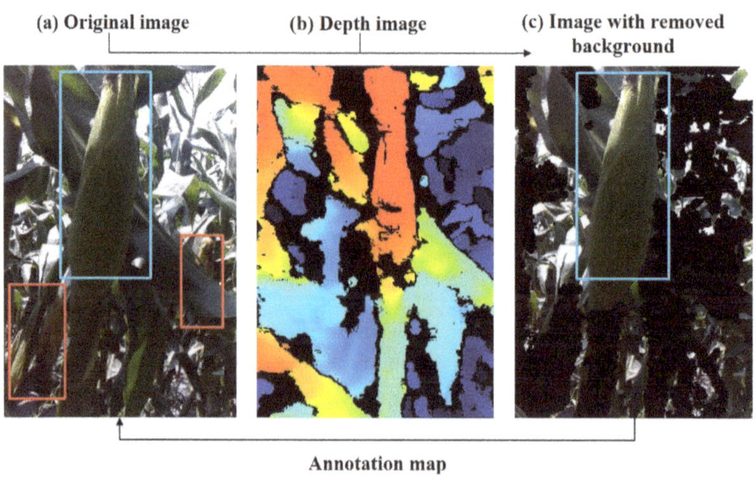

Figure 5. Detection Box Filtering Based on Depth Values. (**a**) Original image. (**b**) Depth image. (**c**) Image with removed background. Red boxes: non-target row ears; blue boxes: target row ears.

https://doi.org/10.3390/plants15010038

In the target-tracking section, t-SNE visualization was applied to features extracted by the original CNN module and the proposed DBC-Net module (Figure 6). The results show that the original CNN has limited discriminative ability for corn ear appearance features in complex field environments, with feature distributions of different ears largely overlapping. In contrast, DBC-Net significantly improves feature separability, producing more compact clusters for the same ear and greater separation between different ears, thereby providing a robust basis for stable tracking using DI-DeepSORT.

The modules function synergistically to form the corn ear height measurement framework. This system not only achieves spatial focus on the target planting rows but also demonstrates higher temporal stability in height estimation. The resulting height measurements are shown in Figure 7, further validating the reliability and practical applicability of the framework in real-world field environments.

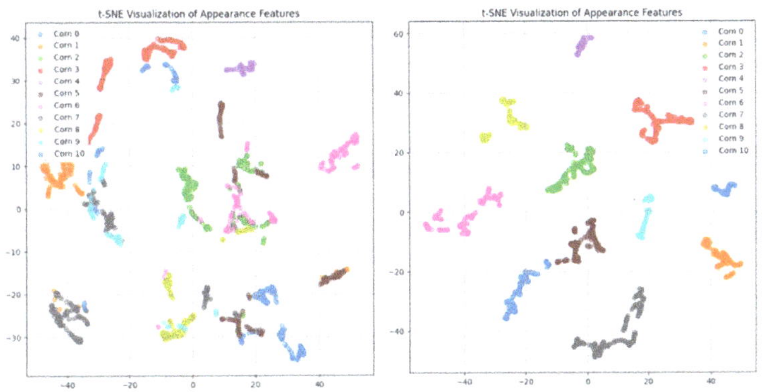

(a) CNN (b) DBC-Net

Figure 6. Visualization of the extracted appearance features of corn ears. (**a**) CNN outputs visualization results of corn ear appearance features (**b**) DBC-Net outputs visualization results of corn ear appearance features.

Figure 7. Height Detection Result. Green boxes: detected targets.

https://doi.org/10.3390/plants15010038

3. Discussion

This study developed a corn ear height measurement model based on YOLO-HAMDF and DI-DeepSORT, effectively addressing challenges in field environments such as severe occlusion, complex backgrounds, and large variations in ear size. The model enhances ear localization accuracy through the TBDA, GLSA, and AQE modules. Depth information is employed to filter non-target regions, and, combined with DI-DeepSORT, stable height sequences are obtained for each ear, providing reliable input for subsequent mechanized operations.

However, the experiments were primarily conducted in fresh corn field scenarios and did not include systematic benchmark testing on large-scale general-purpose datasets such as COCO. Although YOLO-HAMDF demonstrates excellent performance in agriculture-specific tasks, its generalizability remains to be further validated. Future work will conduct benchmark evaluations on public datasets like COCO according to mainstream computer vision standards. Independent ablation studies of the TBDA, GLSA, and LQE modules across tasks and datasets will also be conducted to comprehensively assess their transferability and structural contributions, thereby strengthening the model's novelty and generalizability.

In practical applications, the model's output "height sequences" are not intended for direct point-wise control of harvester parameters. Instead, they are used to construct smoothed and robust dynamic adjustment strategies for the cutter bar. The automatic corn ear detection and height measurement model provides real-time continuous distributions of ear height. By applying sliding-window fusion, trend filtering, and stability evaluation, control signals suitable for the actuator are generated, enabling adaptive adjustment of the cutter bar. This strategy prevents frequent mechanical oscillations while optimizing operational parameters according to real-time field information, demonstrating clear engineering value. Future work will establish a mapping model between corn ear height distributions and cutter bar adjustment parameters, deploying it on edge devices to achieve closed-loop control of harvester operations.

4. Materials and Methods

4.1. FCE-GBD Dataset

The study was conducted at a fresh corn planting site in Shenyang, Liaoning Province, China, as shown in Figure 8A. Image data were collected over eight sessions from July to August 2025, covering different dates, eight independent plots, and various lighting conditions—including sunny, cloudy, and backlit afternoons—to ensure sufficient temporal, spatial, and illumination diversity. RGB and depth images were simultaneously captured using an Intel RealSense D435i depth camera (Intel Corporation, Santa Clara, CA, USA), which operates over a range of 0.2–3 m with a depth measurement error of less than 2% within 2 m. The fresh corn image acquisition platform and its schematic are shown in Figure 8B,C. The platform comprises a field-walking device, the D435i depth camera, and a laptop (Lenovo Group Ltd., Beijing, China), with the camera mounted on a gimbal fixed to the walking device. To simulate actual harvester operating conditions, the camera was installed at a height (h) of 1.8 m, with a pitch angle (α) of 45°, and positioned at a horizontal distance (d) of approximately 30–50 cm from the crop row centerline. The variable H represents the measured height of corn ears under these conditions.

Significant variations in corn plant height and leaf density were observed across the plots. In some plots, dense foliage partially occluded the ears, whereas sparse leaves in other plots left ears more exposed. These variations increase the difficulty of accurate detection. To enhance the generalization and robustness of the model under diverse field growth conditions, images were collected to cover a range of plant heights, ear sizes, and occlusion scenarios, as illustrated in Figure 2. In Figure 9g, ears are fully exposed,

 https://doi.org/10.3390/plants15010038

facilitating detection; in Figure 9h, ears are partially occluded by leaves, increasing detection difficulty; and in Figure 9i, distant ears appear smaller with greater overlap and occlusion, posing further challenges. A total of 4000 raw RGB images and corresponding depth maps were collected, encompassing diverse growth conditions.

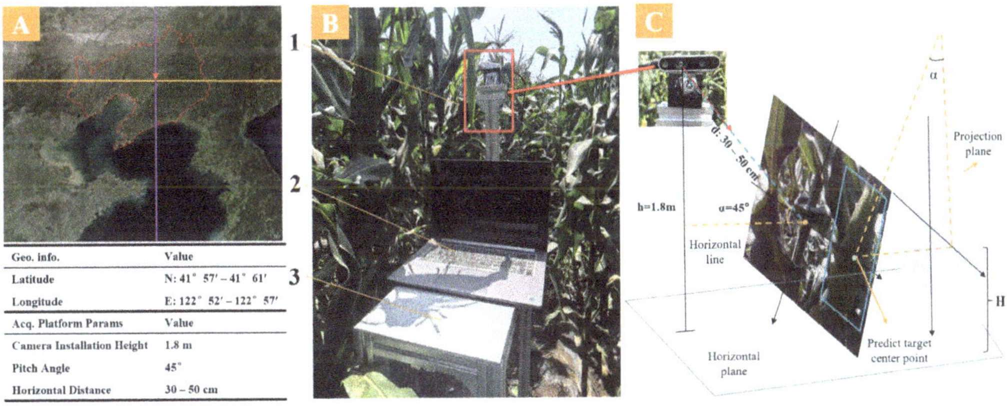

Figure 8. Test area and image acquisition platform. (**A**) Test area. (**B**) Fresh waxy corn image acquisition mobile platform 1. D435i depth camera 2. Computer 3. Mobile car. (**C**) Principle Schematic of Depth Camera.

Figure 9. Images of fresh corn under varying growth conditions. (**a**) Low ear, (**b**) Medium-height ear, (**c**) High ear, (**d**) Small ear, (**e**) Medium-size ear, (**f**) Large ear, (**g**) Fully exposed ear, (**h**) Occluded ear, (**i**) Distant and occluded ear.

https://doi.org/10.3390/plants15010038

All images were manually annotated using LabelImg software (version 1.8.4) to obtain ground-truth positions of fresh corn ears. A standardized annotation protocol was followed: only the visible main regions of ears were labeled; heavily occluded ears that remained partially recognizable were annotated according to their visible portions; objects that were entirely invisible or had minimal visible parts, preventing confirmation of target attributes, were excluded. All annotations were cross-checked by a second annotator to ensure consistency and accuracy of bounding boxes.

The dataset was divided into training, validation, and test sets at an 8:1:1 ratio, containing 3200, 400, and 400 images, respectively. Splitting strictly adhered to a sequence-independent principle, ensuring that images from the same plot or captured consecutively on the same day did not appear in different subsets, thereby preventing data leakage. The resulting dataset was named the FCE-GBD Dataset.

4.2. CEHD

The framework of the proposed corn ear height detection (CEHD) is illustrated in Figure 10. Images are first input into the YOLO-HAMDF network for ear detection, and non-target ears in adjacent rows are filtered using depth maps to reduce false positives. The filtered bounding boxes are then passed to the DI-DeepSORT module to achieve continuous and stable cross-frame tracking of individual ears. For each tracked target, the CHSM-PCM module computes the local point cloud by back-projecting depth pixels within the predicted bounding box, removing outliers, and calculating the median height. Combined with the camera's extrinsic parameters, the absolute height of the ear relative to the ground is obtained. Through the coordinated processes of detection, tracking, and height estimation, the framework enables high-precision recognition and reliable height measurement of corn ears under complex field conditions.

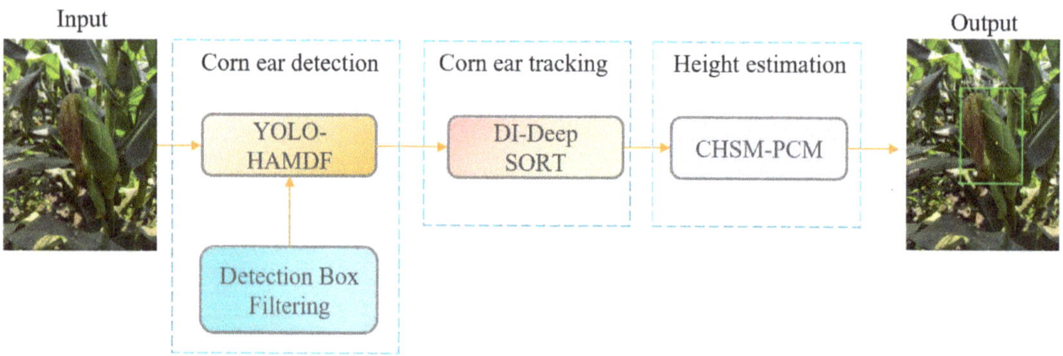

Figure 10. The architecture of CEHD. Green boxes: detected targets.

4.3. YOLO-HAMDF

4.3.1. YOLO-HAMDF Architecture

The network framework was designed based on YOLO11-L with several efficient module innovations, resulting in the proposed YOLO-HAMDF network. The overall architecture is illustrated in Figure 11. An innovative TBDA module was introduced to replace the original C3 module in the backbone. By combining multi-scale convolutional branches with channel attention, this module effectively enhances the capture of ear details at different scales. To further strengthen feature representation, a GLSA module was incorporated into the Neck, replacing the original YOLO11-L Concat + C3k2 modules with Fusion + Node-Mode + GLSA modules, which were stacked three times at each scale. Finally, the traditional detection heads were replaced with the AQE module to better integrate with depth camera data for corn ear height estimation. The detailed construction of the network is described in the following sections.

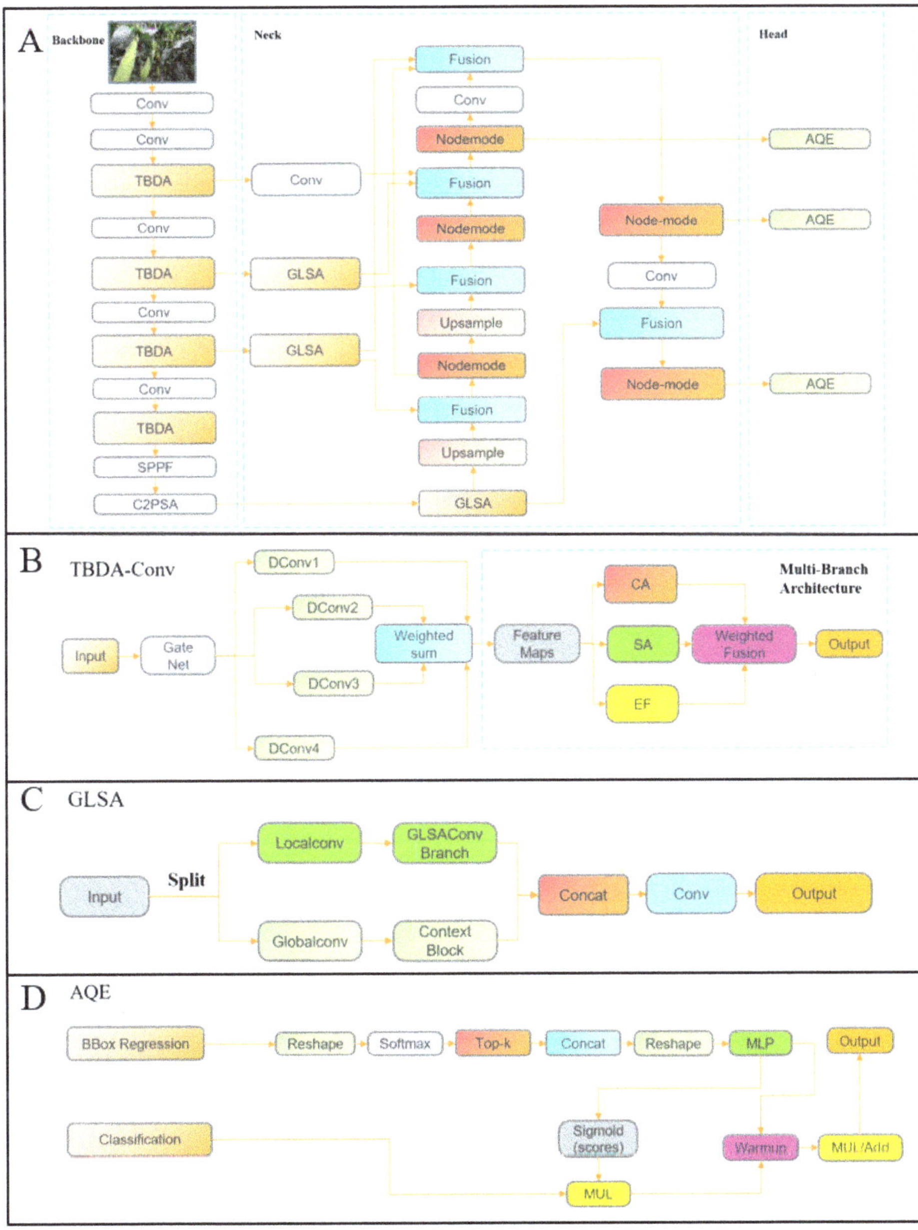

Figure 11. Architecture of the proposed YOLO-HAMDF. (**A**) Overall architecture. (**B**) Architecture of TBDA. (**C**) Architecture of GLSA. (**D**) Architecture of AQE.

4.3.2. TBDA

Attention mechanisms allow networks to focus on critical features, thereby enhancing object recognition. To address challenges posed by complex background interference in corn ear height detection, the TBDA module was designed. It integrates multiple attention

 https://doi.org/10.3390/plants15010038

mechanisms to strengthen feature representation and improve detection accuracy, as shown in Figure 12.

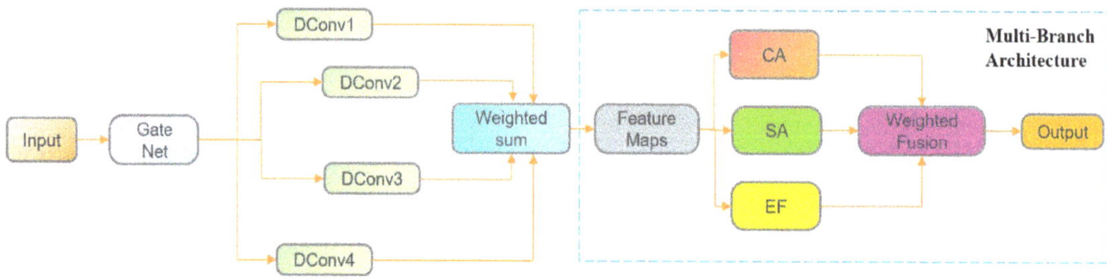

Figure 12. Architecture of TBDA.

The module first introduces a dynamic convolution structure based on a gating mechanism to enable adaptive modeling of multi-scale features. It contains four convolutional branches with kernel sizes of 1×1, 3×3, 5×5, and 7×7. A lightweight gating network dynamically generates weight coefficients for each branch. Specifically, global average pooling is applied to the input feature map to extract contextual information, followed by a 1×1 convolution and a Sigmoid function to produce branch-specific weight responses. The convolution outputs of all branches are then combined via weighted fusion along the channel dimension, allowing selective enhancement across different receptive fields.

A three-branch attention structure further strengthens feature representation. The first branch uses a channel attention (CA) module to emphasize important semantic channels and suppress irrelevant or noisy channels, enabling the model to focus on spectral and textural features critical for corn ears. The second branch employs a spatial attention (SA) module to highlight regions with strong responses, reinforcing target areas and suppressing background interference, thereby improving localization precision and spatial sensitivity. The third branch applies an energy function (EF) to generate spatial weights based on statistical differences, adaptively amplifying locally salient features and enhancing discriminability under complex backgrounds. Finally, the outputs of the three branches are fused using learnable parameters, providing complementary enhancement across channel, spatial, and energy information and further improving detection performance.

4.3.3. GLSA

From an overall perspective, although the backbone integrates the innovative TBDA convolution with efficient modules such as SPPF and C2PSA to enhance high-level semantic feature representation and suppress background interference, it still struggles to capture long-range dependencies. In complex field scenarios, the network remains susceptible to background distractions, occlusions, and target deformations. To address this issue, the Global-Local Selective Aggregation (GLSA) module was introduced (Figure 13) to enhance feature representation at the Neck stage, thereby improving the localization and recognition accuracy of corn ears. The GLSA module is deployed at the forefront of three detection branches—P3 (low-to-mid level), P4 (mid-to-high level), and P5 (high level)—so that each branch receives enhanced GL features before entering the fusion and decoding paths, effectively improving the perceptual and discriminative capacity of the base features.

Within the GLSA module, the input feature channels are divided into two branches: the Local branch employs a lightweight MLP-style GLSA Conv Branch to model fine-grained features such as local textures and edges; the Global branch leverages a context block based on contextual attention to aggregate global contextual information, complementing the spatial-structural semantics of the target. The outputs of the two branches are then

 https://doi.org/10.3390/plants15010038

concatenated along the channel dimension and integrated through a convolution to produce a unified output.

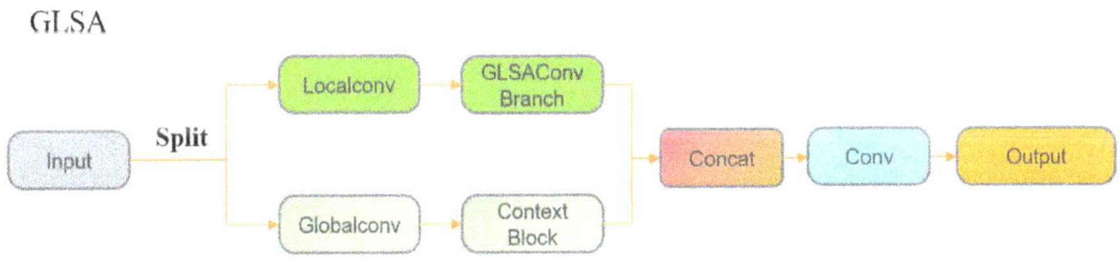

Figure 13. Architecture of GLSA.

The semantic features enhanced by the GLSA module provide stable support for the subsequent LQE, which is based on boundary quality estimation, thereby improving the accuracy and reliability of predicted bounding boxes. By serving as a bridge for global–local information interaction between the backbone and detection heads, the GLSA module demonstrates strong adaptability and robustness in real-world field detection scenarios.

4.3.4. AQE

In recent years, LQE had been widely adopted in the field of object detection. It statistically analyzed the probability distribution of bounding box regression, extracted key distribution features, and then employed a multi-layer perceptron to predict a quality score, which was used to adjust and optimize the bounding box ranking in object detection. This approach effectively improved the accuracy and reliability of the predicted bounding boxes. The specific steps were as follows:

First, the predicted bounding box corner point distribution $P \in R^{B\times(4\times reg)\times H\times W}$ was normalized using *softmax* to obtain the probability distribution *prob* of each regression category at every location.

$$prob = softmax(P) \tag{1}$$

From the normalized probability distribution, the top K highest probabilities are selected, and their mean was computed to serve as the statistical feature *stat* for that location.

$$stat = [top - k(prob), mean(top - k(prob))] \tag{2}$$

The statistical feature was then fed into a multilayer perceptron (*MLP*) to generate a single-channel quality score Q, which reflected the confidence of the bounding box prediction at that location.

$$Q = MLP(stat) \tag{3}$$

Finally, the LQE integrated the quality score with the initial classification confidence to refine the final bounding box score S', thereby providing a more accurate measure of detection quality.

$$S' = S + Q \tag{4}$$

Through the above process, the LQE not only relied on classification scores but also incorporated statistical information from the regression distribution, thereby effectively improving localization accuracy. However, the original LQE model presented a potential issue when dealing with partially occluded or distant fresh corn ears: such hard-to-classify targets could receive high bounding box quality scores due to relatively accurate localization, even when their classification confidence was insufficient. As a result, these unreliable boxes might still be output by the detector, leading to false detections that not only reduced overall detection accuracy but also hindered subsequent corn height estimation. To address

https://doi.org/10.3390/plants15010038

this issue, the LQE module was further enhanced to propose the AQE, with the module architecture shown in Figure 14.

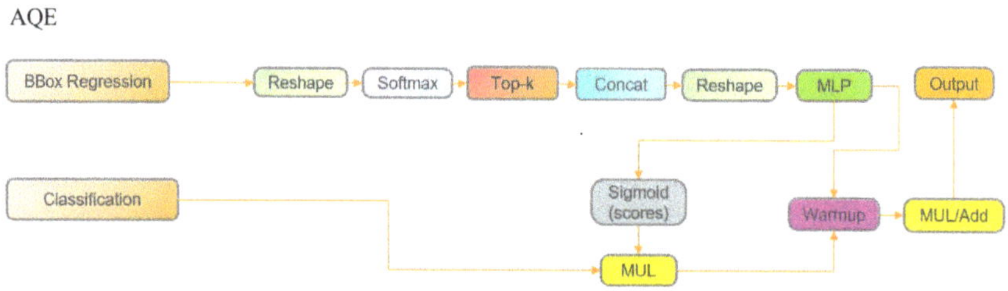

Figure 14. Architecture of AQE.

First, an occlusion factor O was introduced, which was derived from the initial classification score S through a Sigmoid activation:

$$O = \sigma(S) \tag{5}$$

This factor was employed to adjust the confidence of the bounding box. In cases where the classification score was low but the box quality score was high, the final score of the detection box was reduced to support subsequent height estimation. Specifically, when the detection score did not meet the requirement, the box was temporarily suppressed until occlusion decreased and the classification score increased as the machine advanced. The detection box was finally output only when the overall score S' reached the threshold (empirically $S' \geq 0.75$).

In addition, a dynamic adjustment strategy was introduced during training. In the early training stage, an additive approach was adopted to encourage the model to learn as much comprehensive information as possible.

$$S' = S + Q \times O \tag{6}$$

In the later stage, after the model had converged, a multiplicative fusion strategy was adopted to more strictly suppress detection responses from low-quality regions.

$$S' = S \times \sigma(Q \times O) \tag{7}$$

This strategy enabled the model to suppress output when detection performance was suboptimal, and to release the corresponding prediction boxes only after the confidence had gradually increased, thereby effectively avoiding overestimation caused by low-quality boxes and enhancing the robustness and reliability of corn ear height detection.

4.4. DI-DeepSORT

4.4.1. DI-DeepSORT Architecture

In this study, the traditional DeepSORT tracking framework was systematically enhanced, resulting in the proposed DI-DeepSORT (Figure 15). First, to improve the representation of object re-identification (ReID) features, the DBC-Net was designed. This network, based on a dual-line convolution structure combined with a channel–spatial attention mechanism (CBAM), more effectively captures the spatial–spectral features and local details of targets, thereby enhancing the discriminability of multi-object features. Second, to address the sensitivity to noise and position jumps in conventional Kalman filtering for state estimation, the Independent-Dimension Adaptive Kalman Filter (IDA-Kalman) was introduced. This filter adaptively adjusts measurement noise along independent dimen-

https://doi.org/10.3390/plants15010038

sions and applies smoothing to the position center, ensuring stable trajectory estimation under complex conditions. The combined improvements significantly enhance the accuracy and robustness of multi-object tracking with the modified DeepSORT in challenging scenarios.

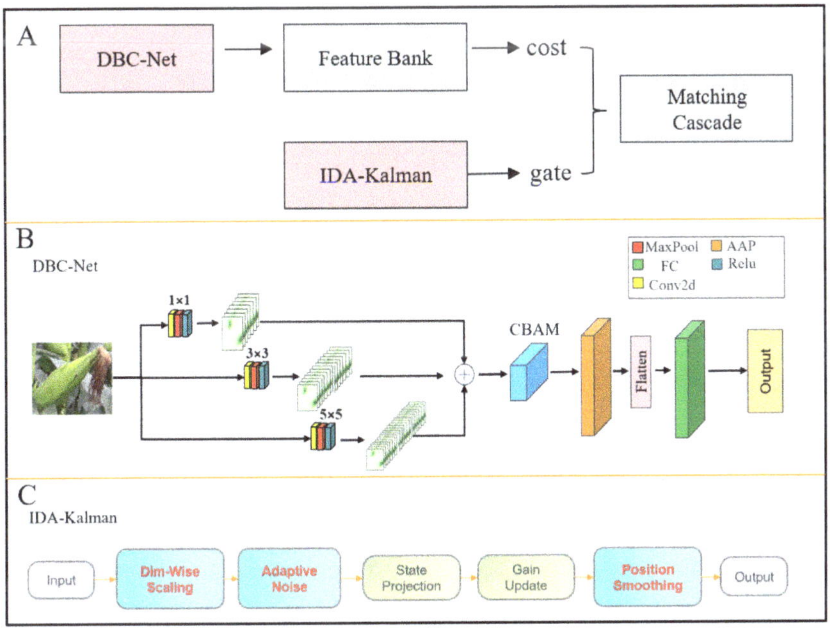

Figure 15. Architecture of the proposed DI-DeepSORT. (**A**) Overall architecture. (**B**) Architecture of DBC-Net. (**C**) Architecture of IDA-Kalman.

4.4.2. DBC-Net

The DBC-Net architecture is illustrated in Figure 16. In the backbone branch, three convolution operations are sequentially applied with kernel sizes of 1, 3, and 5, and corresponding padding sizes of 0, 1, and 2. Convolutional kernels of different scales extract features from multiple receptive fields: the 1 × 1 convolution primarily compresses and fuses channel information, the 3 × 3 convolution captures fine-grained local features, and the 5 × 5 convolution encodes broader contextual information. This multi-scale convolutional design enables the network to preserve local textural details of corn ears while enhancing its perception of the overall structure. The features obtained from the three convolutional branches are subsequently fused and fed into the CBAM module. Within CBAM, channel and spatial attention mechanisms are used to weight the features. Adaptive average pooling (AAP), a flattening layer, and a fully connected (FC) layer are applied to generate the final feature representation. By integrating local fine-grained details with global structural information, the network achieves a more comprehensive representation of corn ear features.

4.4.3. IDA-Kalman

In the conventional DeepSORT framework, the Kalman filter was used to predict object states and update trajectories, with the target state vector defined as: $x_k = \left[x, y, a, h, \dot{x}, \dot{y}, \dot{a}, \dot{h}\right]^T$.

Here, (x,y) represents the target center position, a denotes the width-to-height ratio, h is the target height, and $\dot{x}, \dot{y}, \dot{a}, \dot{h}$ corresponds to the respective velocities. During the

 https://doi.org/10.3390/plants15010038

prediction stage, the target state in the current frame was estimated by the Kalman filter based on the state from the previous frame:

$$x_{k|k-1} = Fx_{k-1|k-1} + B_{uk} \tag{8}$$

$$P_{k|k-1} = FP_{k-1|k-1}F^T + Q \tag{9}$$

F is the state transition matrix, B is the control matrix, and Q is the process noise covariance matrix. When the detector provided new observations, the Kalman filter used these measurements to correct the predicted state:

$$K_k = P_{k|k-1}H^T\left(HP_{k|k-1}H^T + R\right)^{-1} \tag{10}$$

$$x_{k|k} = x_{k|k-1} + K_k\left(z_k - Hx_{k|k-1}\right) \tag{11}$$

$$P_{k|k=}(I - K_kH)P_{k|k-1} \tag{12}$$

H is the observation matrix, R is the observation noise covariance matrix, and z_k represents the observation provided by the detector. In conventional Kalman filtering, Q and R are typically set as fixed values. However, in practical object detection scenarios, the quality of detector outputs may vary significantly due to changes in illumination, occlusion, and target scale. A fixed R cannot adequately reflect the uncertainty of observations, causing the filter to over-rely on noisy measurements when detection confidence is low, which may result to bounding box instability.

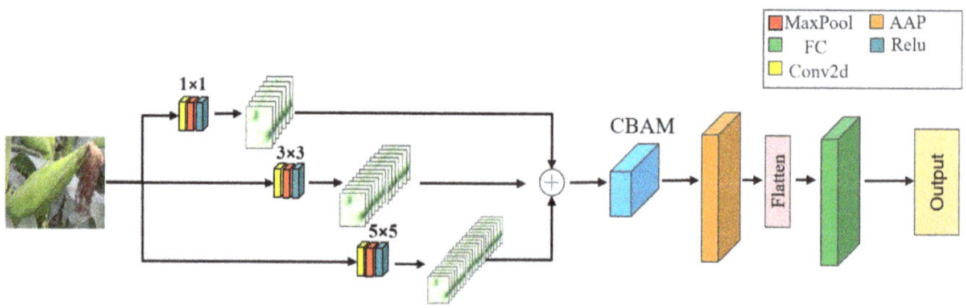

Figure 16. The structure diagram of the DBC-Net in this study.

To address the insufficient targeted modeling of measurement noise in conventional Kalman filtering, the Independent-Dimension Adaptive Kalman Filter (IDA-Kalman) was proposed in this study, as illustrated in Figure 17.

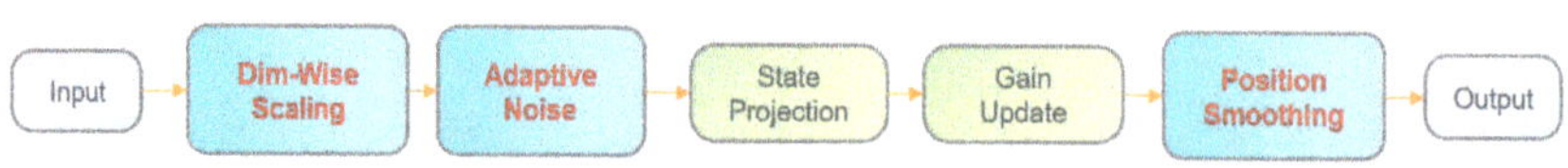

Figure 17. The structure diagram of the IDA-Kalman in this study.

This method introduces improvements in measurement noise modeling and position update strategies to enhance the robustness of multi-object tracking in complex scenarios. The observation noise covariance matrix in IDA Kalman was redefined as:

$$R_t = \text{diag}\left(\sigma_x^2, \sigma_y^2, \sigma_a^2, \sigma_h^2\right) \tag{13}$$

https://doi.org/10.3390/plants15010038

Here, the measurement noise σ_i for each dimension is dynamically adjusted using the weighted detection confidence $c \in [0, 1]$:

$$\begin{aligned} \sigma_x &= max(1 - c, \varepsilon) \cdot w \cdot h_{t-1} \\ \sigma_y &= max(1 - 0.8c, \varepsilon) \cdot w \cdot h_{t-1} \\ \sigma_a &= max(1 - 0.5c, \varepsilon) \cdot 0.1 \\ \sigma_h &= max(1 - c, \varepsilon) \cdot w \cdot h_{t-1} \end{aligned} \tag{14}$$

h_{t-1} represents the target height in the previous frame, w is the position normalization weight, and $\varepsilon = 0.01$ denotes the noise lower bound, preventing the noise from becoming too small as c approaches 1. In this study, the measurement noise weights are individually defined for the four state dimensions based on empirical detection performance. Compared with the fixed R in conventional Kalman filtering, IDA Kalman can independently adjust the noise magnitude according to detection confidence and dimension-specific weights, relying more on observations when detection is reliable and more on predictions when detection is uncertain, thereby enhancing the filter's adaptability to uncertainty. The improved Kalman gain formula is expressed as:

$$K_t = P_{t|t-1} H^T \left(H P_{t|t-1} H^T + R_t \right)^{-1} \tag{15}$$

$$x_t' = x_{t|t-1} + K_t \left(z_t - H x_{t|t-1} \right) \tag{16}$$

R_t has already been dynamically adjusted according to the detection confidence. In addition, to further suppress the jitter of the center position (x,y), an exponential smoothing mechanism is introduced. After incorporating this mechanism, the expression for x_t is as follows, and y_t is treated similarly:

$$x_t = \alpha x_{t-1} + (1 - \alpha) x_t' \tag{17}$$

$\alpha \in [0, 1]$ represents the smoothing coefficient, with larger values indicating a more stable position.

By employing independent-dimension adaptive noise modeling and smoothing updates of the center position, the filter's adaptability to observation uncertainty is significantly enhanced, effectively improving the stability and accuracy of multi-object tracking in complex environments.

4.5. CHSM-PCM

After achieving continuous and stable tracking of corn ears, the Corn Height Solving Module based on Point-Cloud Median (CHSM-PCM) was applied to each tracked target to estimate its height, as illustrated in Figure 18. The predicted bounding boxes obtained from tracking were used as input, and the corresponding depth pixels within each box were back-projected to form a local point cloud. Depth outliers were removed, and the spatial height of each corn ear relative to the camera was computed as the median along the height axis of the point cloud. By incorporating the camera's extrinsic parameters, including installation height and tilt angle, the absolute height of each corn ear relative to the ground was obtained, enabling reliable height estimation for all target ears in each frame.

4.6. Experimental Settings

4.6.1. Parameter Settings

In this study, the deep learning hardware platform for the fresh corn ear detection task consisted of a laptop equipped with an Intel Core i7-11700 CPU (8 cores, Intel Corporation,

https://doi.org/10.3390/plants15010038

Santa Clara, CA, USA), 16 GB of RAM, and an NVIDIA GeForce GTX 4060 GPU (16 GB VRAM, 3584 CUDA cores, NVIDIA Corporation, Santa Clara, CA, USA). A single GPU was used during training, with an average iteration time of approximately 30 s per epoch. The software environment included Python 3.8.8, PyTorch 1.9.0, torchvision 0.10.0, CUDA 11.1, and OpenCV-python 4.5.3.

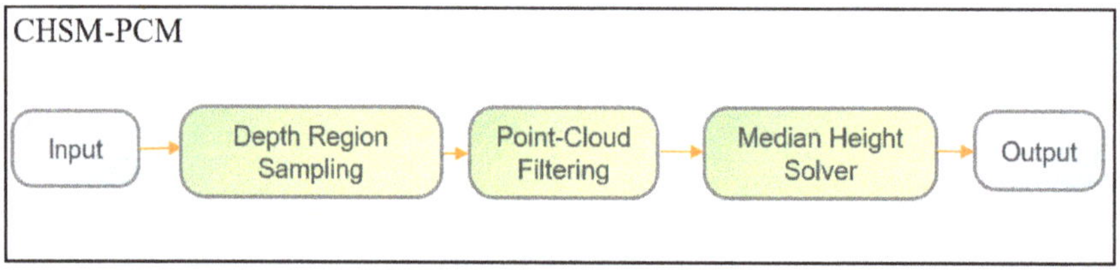

Figure 18. Architecture of the proposed CHSM-PCM.

The network input size was set to 640 × 640 pixels, with a batch size of 4 and a total of 300 training iterations. Stochastic gradient descent (SGD) served as the optimizer, with a momentum of 0.937, weight decay of 5×10^{-4}, and an initial learning rate of 0.01. A cosine annealing schedule was applied to dynamically adjust the learning rate and improve training stability. During training, data augmentation was performed, including random horizontal flipping, rotation (±15°), brightness/contrast adjustment, random cropping, and HSV color space perturbation, to enhance model generalization. All models used for corn ear detection were trained on the same dataset with identical parameters to ensure fairness and consistency in comparative evaluations.

4.6.2. Evaluation Metrics

In the field of fresh maize ear detection, commonly used performance evaluation metrics include IoU (Intersection over Union), Precision, Recall, and mAP (mean Average Precision). In object detection tasks, a target is generally considered successfully detected when IoU exceeds 0.5; therefore, an IoU threshold of 0.5 was adopted in this study. True Positives (*TP*) are defined as the number of positive samples correctly identified as maize ears, whereas False Positives (*FP*) denote the number of negative samples incorrectly classified as maize ears. True Negative (*TN*) denotes the number of negative samples correctly recognized as non-maize ears, and False Negative (*FN*) represents the number of actual maize ears that were not detected. *Precision* is calculated as the proportion of correctly identified maize ears among all predicted maize ears, while *Recall* is calculated as the proportion of correctly identified maize ears among all actual maize ears. The formulas for these metrics are expressed as follows:

$$Precision = \frac{TP}{TP + FP} \tag{18}$$

$$Recall = \frac{TP}{TP + FN} \tag{19}$$

In this study, model performance is comprehensively evaluated using multi-scale and multi-threshold mean Average Precision (mAP) metrics, including mAP@[0.5: 0.95], mAP@0.5, and mAP@0.75, along with average precision (AP) for targets of different sizes according to the COCO standard. During mAP calculation, the confidence threshold for predicted boxes is set to 0.25, and non-maximum suppression (NMS) with a threshold of 0.45 is applied to remove duplicate detections.

Furthermore, to comprehensively assess the accuracy of the corn ear height detection model integrating object detection with depth information, mean absolute error (MAE) and

 https://doi.org/10.3390/plants15010038

the Pearson correlation coefficient (r) are employed as evaluation metrics. MAE provides an intuitive measure of the average deviation between predicted and manually measured heights, expressed in the same units as corn ear height, thus serving as a suitable metric for evaluating height measurement errors. Manual measurements were performed using a steel tape measure (minimum scale 1 mm), from the ground at the center of the inter-row space to the midpoint of each ear. Each sample was independently measured twice by two experimenters, and the mean of the two measurements was used. The standard deviation of repeated measurements was less than 0.5 cm, confirming the reliability of the ground-truth heights. The Pearson correlation coefficient quantifies the linear relationship between automated measurements and manual references, reflecting the consistency and stability of the predicted trends. This metric complements MAE by providing insight into trend agreement, thereby enabling a more comprehensive evaluation of model performance.

The specific formulas are as follows:

$$MAE = \frac{1}{N}\sum_{i=1}^{N}|\hat{y}_i - y_i| \tag{20}$$

where $\hat{y}_i$ denotes the machine-measured height of the $i-th$ sample, y_i represents the corresponding manually measured height, and N is the total number of samples.

$$r = \frac{\sum_{i=1}^{N}(\hat{y}_i - \hat{y})(y_i - \overline{y})}{\sqrt{\sum_{i=1}^{N}(\hat{y}_i - \overline{\hat{y}})^2}\sqrt{\sum_{i=1}^{N}(y_i - \overline{y})^2}} \tag{21}$$

$\overline{\hat{y}}$ and $\overline{y}$ denote the mean values of the machine-measured and manually measured heights, respectively, and r ranges from -1 to 1, with values closer to 1 indicating a stronger linear correlation between the two sets of measurements.

5. Conclusions

This study integrates RGB and depth image data to address the challenges of corn ear detection in complex fresh corn field environments and proposes a corn ear height measurement model based on YOLO-HAMDF and DI-DeepSORT. Compared with conventional detection methods, the model employs the innovatively designed TBDA multi-scale dynamic convolution module, GLSA global-local feature aggregation module, and the improved AQE detection head, integrated with the DI-DeepSORT tracking module, thereby significantly enhancing both ear recognition accuracy and height measurement precision. The main conclusions are as follows:

(1) The synergistic integration of TBDA, GLSA, and AQE allows the model to achieve enhanced feature representation and discriminative capability in complex backgrounds and densely distributed corn ear scenarios. Under the combined effect of these modules, the model attains a detection accuracy of 97.01%, outperforming other state-of-the-art methods.

(2) By incorporating depth information from the Intel RealSense camera, non-target row ears can be effectively filtered, thereby reducing interference from non-target ears and enhancing both computational efficiency and practical applicability of the system.

(3) The integration of the IDA Kalman filter with the DBC-Net feature extraction network facilitates more stable cross-frame matching. When paired with the YOLO-HAMDF model for height measurement, the MAE is reduced to 3.21 ± 0.05 cm, and the Pearson correlation coefficient increases to 0.92 ± 0.01. This effectively mitigates detection box jitter caused by occlusion or detection fluctuations, providing continuous and reliable

https://doi.org/10.3390/plants15010038

input for height curve extraction, further enhancing the overall stability and accuracy of the measurement.

Author Contributions: H.W.: Writing—review and editing, Writing—original draft, Data curation, Formal analysis, Conceptualization, Software. Y.L.: Writing—review and editing, Software, Validation. J.F.: Project administration, Resources, Funding acquisition. Q.F.: Validation, Visualization, Supervision, Data curation. Y.Q.: Methodology, Visualization, Supervision. All authors have read and agreed to the published version of the manuscript.

Funding: This work was supported by the National Key R&D Program of China (Grant No. 2025YFE0103400), the National Natural Science Foundation of China (Grant No. 52575287) and the Scientific Research Project of Education Department of Jilin Province (Grant No. JJKH20241316HT).

Data Availability Statement: The datasets generated and analyzed during the current study, including RGB images, depth maps, and annotation files of fresh corn ears (FCE-GBD Dataset), are publicly available at [https://github.com/why110054/corn] (accessed on 31 October 2025). The dataset is released for academic research purposes under a CC BY 4.0 license. Researchers can freely download and use the data, provided that appropriate citation to the original study is given.

Acknowledgments: The authors would like to thank all individuals who contributed to this work. All individuals mentioned in this section have given their consent to be acknowledged.

Conflicts of Interest: The authors declare that they have no known competing financial interests or personal relationships that could have appeared to influence the work reported in this paper.

Abbreviations

The following abbreviations are used in this manuscript:

RGB	Red, Green, Blue
CNN	Convolutional Neural Network
mAP	Mean Average Precision
IoU	Intersection over Union
MAE	Mean Absolute Error
TBDA	Three-Branch Dynamic Attention module
GLSA	Global-Local Selective Aggregation module
LQE	Location Quality Estimator
AQE	Adaptive Quality Estimator
DI-DeepSORT	DBC-IDA DeepSORT
DBC-Net	Dual-Branch Convolutional Network
IDA-Kalman	Independent-Dimension Adaptive Kalman Filter
CHSM-PCM	Corn Height Solving Module based on Point-Cloud Median
FPS	Frames per Second
CEHD	Corn Ear Height Detection framework
FCE-GBD Dataset	Fresh Corn Ear—Ground-Based RGB-D Dataset

References

1. Karanam, B.T.; Kumari, V.N.; Sivakumar, S.; Vanitha, K.; Meenakshi, P.; Himakara, D.M.; Rawat, P. Nutritional enhancement and genetic innovations in sweet corn: Unlocking super sweetness and health benefits through modern breeding technique—A review. *Plant Sci. Today* **2024**, *11*, 5441. [CrossRef]
2. Wang, H.; Liu, J.S.; Min, W.H.; Zheng, M.Z.; Li, H. Changes of moisture distribution and migration in fresh ear corn during storage. *J. Integr. Agric.* **2019**, *18*, 2644–2651. [CrossRef]
3. Guan, X.; Li, T.; Zhou, F. Determination of bruise susceptibility of fresh corn to impact load by means of finite element method simulation. *Postharvest Biol. Technol.* **2023**, *198*, 112227. [CrossRef]
4. Li, Y.; Fu, J.; Fu, Q.; Wang, H.; Chen, Z.; Liu, X. Finite element analysis of the compression behavior and damage mechanisms of fresh corn kernels. *Ind. Crops Prod.* **2025**, *233*, 121453. [CrossRef]

https://doi.org/10.3390/plants15010038

5. Bissadu, K.D.; Sonko, S.; Hossain, G. Society 5.0 enabled agriculture: Drivers, enabling technologies, architectures, opportunities, and challenges. *Inf. Process. Agric.* **2025**, *12*, 112–124. [CrossRef]
6. Wang, Q.; Meng, Z.-J.; Wen, C.-K.; Qin, W.-C.; Wang, F.; Zhang, A.-Q.; Zhao, C.-J.; Yin, Y.-X. Grain combine harvester header profiling control system development and testing. *Comput. Electron. Agric.* **2024**, *223*, 109082. [CrossRef]
7. Xiao, C.; Zheng, L.; Li, M.; Chen, Y.; Mai, C. Apple detection from apple tree image based on BP neural network and Hough transform. *Int. J. Agric. Biol. Eng.* **2015**, *8*, 46–53. [CrossRef]
8. Wachs, J.P.; Stern, H.I.; Burks, T.; Alchanatis, V. Low and high-level visual feature-based apple detection from multi-modal images. *Precis. Agric.* **2010**, *11*, 717–735. [CrossRef]
9. Lan, H.; Jin, S. An Improved Suppressed FCM Algorithm for Image Segmentation. In Proceedings of the 4th International Conference on Manufacturing Science and Engineering (ICMSE 2013), Dalian, China, 30–31 March 2013; pp. 2349–2353.
10. Keramat-Jahromi, M.; Mohtasebi, S.S.; Mousazadeh, H.; Ghasemi-Varnamkhasti, M.; Rahimi-Movassagh, M. Real-time moisture ratio study of drying date fruit chips based on on-line image attributes using kNN and random forest regression methods. *Measurement* **2021**, *172*, 108899. [CrossRef]
11. Zhang, Y.; Wu, L. Classification of Fruits Using Computer Vision and a Multiclass Support Vector Machine. *Sensors* **2012**, *12*, 12489–12505. [CrossRef]
12. Dosovitskiy, A. An image is worth 16 × 16 words: Transformers for image recognition at scale. *arXiv* **2020**, arXiv:2010.11929.
13. Liu, Z.; Lin, Y.; Cao, Y.; Hu, H.; Wei, Y.; Zhang, Z.; Lin, S.; Guo, B. Swin transformer: Hierarchical vision transformer using shifted windows. In Proceedings of the IEEE/CVF International Conference on Computer Vision, Montreal, QC, Canada, 10–17 October 2021; pp. 10012–10022.
14. Zhou, Q.; Huang, Z.L.; Zheng, S.J.; Jiao, L.; Wang, L.S.; Wang, R.J. A wheat spike detection method based on Transformer. *Front. Plant Sci.* **2022**, *13*, 1023924. [CrossRef] [PubMed]
15. Wu, Y.L.; Yuan, S.Q.; Tang, Y.; Tang, L.D. Application of real-time detection transformer based on convolutional block attention module and grouped convolution in maize seedling. *Front. Plant Sci.* **2025**, *16*, 1672746. [CrossRef] [PubMed]
16. Bagci, R.S.; Acar, E.; Türk, Ö. Identification of cotton and corn plant areas by employing deep transformer encoder approach and different time series satellite images: A case study in Diyarbakir, Turkey. *Comput. Electron. Agric.* **2023**, *209*, 107838. [CrossRef]
17. Zhang, N.; Chen, Y.Q.; Zhang, E.X.; Liu, Z.Y.; Yue, J. Maize quality detection based on MConv-SwinT high-precision model. *PLoS ONE* **2025**, *20*, e0312363. [CrossRef]
18. Zhu, Z.B.; Gao, Z.K.; Zhuang, J.J.; Huang, D.C.; Huang, G.G.; Wang, H.S.; Pei, J.W.; Zheng, J.J.; Liu, C.Y. MSMT-RTDETR: A Multi-Scale Model for Detecting Maize Tassels in UAV Images with Complex Field Backgrounds. *Agriculture* **2025**, *15*, 1653. [CrossRef]
19. Parez, S.; Dilshad, N.; Lee, J.W. A Channel Attention-Driven Optimized CNN for Efficient Early Detection of Plant Diseases in Resource Constrained Environment. *Agriculture* **2025**, *15*, 127. [CrossRef]
20. Gai, R.; Chen, N.; Yuan, H. A detection algorithm for cherry fruits based on the improved YOLO-v4 model. *Neural Comput. Appl.* **2023**, *35*, 13895–13906. [CrossRef]
21. Chen, Z.; Chen, B.; Huang, Y.; Zhou, Z. GE-YOLO for Weed Detection in Rice Paddy Fields. *Appl. Sci.* **2025**, *15*, 2823. [CrossRef]
22. Zhou, Q.; Li, H.; Cai, Z.; Zhong, Y.; Zhong, F.; Lin, X.; Wang, L. YOLO-ACE: Enhancing YOLO with Augmented Contextual Efficiency for Precision Cotton Weed Detection. *Sensors* **2025**, *25*, 1635. [CrossRef]
23. Sun, J.; Peng, Y.; Chen, C.; Zhang, B.; Wu, Z.; Jia, Y.; Shi, L. ESC-YOLO: Optimizing apple fruit recognition with efficient spatial and channel features in YOLOX. *J. Real-Time Image Process.* **2024**, *21*, 162. [CrossRef]
24. Zheng, S.; Liu, Y.; Weng, W.; Jia, X.; Yu, S.; Wu, Z. Tomato Recognition and Localization Method Based on Improved YOLOv5n-seg Model and Binocular Stereo Vision. *Agronomy* **2023**, *13*, 2339. [CrossRef]
25. Zhong, S.; Xu, W.; Zhang, T.; Chen, H. Identification and Depth Localization of Clustered Pod Pepper Based on Improved Faster R-CNN. *IEEE Access* **2022**, *10*, 93615–93625. [CrossRef]
26. Li, Y.; He, L.; Jia, J.; Lv, J.; Chen, J.; Qiao, X.; Wu, C. In-field tea shoot detection and 3D localization using an RGB-D camera. *Comput. Electron. Agric.* **2021**, *185*, 106149. [CrossRef]
27. Parr, B.; Legg, M.; Alam, F. Grape yield estimation with a smartphone's colour and depth cameras using machine learning and computer vision techniques. *Comput. Electron. Agric.* **2023**, *213*, 108174. [CrossRef]
28. Zhao, Y.; Zhang, X.; Sun, J.; Yu, T.; Cai, Z.; Zhang, Z.; Mao, H. Low-Cost Lettuce Height Measurement Based on Depth Vision and Lightweight Instance Segmentation Model. *Agriculture* **2024**, *14*, 1596. [CrossRef]
29. Huang, C.; Zeng, Q.; Xiong, F.; Xu, J. Space dynamic target tracking method based on five-frame difference and Deepsort. *Sci. Rep.* **2024**, *14*, 6020. [CrossRef]
30. Wang, Z.; Walsh, K.; Koirala, A. Mango Fruit Load Estimation Using a Video Based MangoYOLO-Kalman Filter-Hungarian Algorithm Method. *Sensors* **2019**, *19*, 2742. [CrossRef]
31. Li, R.; Liu, Q.M.; Wang, M.; Su, Y.C.; Li, C.; Ou, M.X.; Liu, L. Maize Kernel Batch Counting System Based on YOLOv8-ByteTrack. *Sensors* **2025**, *25*, 5584. [CrossRef]

 https://doi.org/10.3390/plants15010038

32. Ye, X.Q.; Pan, J.; Shao, F.; Liu, G.S.; Lin, J.Y.; Xu, D.X.; Liu, J. Exploring the potential of visual tracking and counting for trees infected with pine wilt disease based on improved YOLOv5 and StrongSORT algorithm. *Comput. Electron. Agric.* **2024**, *218*, 108671. [CrossRef]
33. Tu, S.Q.; Huang, Y.F.; Liang, Y.; Liu, H.X.; Cai, Y.F.; Lei, H. A passion fruit counting method based on the lightweight YOLOv5s and improved DeepSORT. *Precis. Agric.* **2024**, *25*, 1731–1750. [CrossRef]
34. Aharon, N.; Orfaig, R.; Bobrovsky, B.-Z. BoT-SORT: Robust associations multi-pedestrian tracking. *arXiv* **2022**, arXiv:2206.14651.
35. Ren, S.; He, K.; Girshick, R.; Sun, J. Faster R-CNN: Towards Real-Time Object Detection with Region Proposal Networks. In Proceedings of the 29th Annual Conference on Neural Information Processing Systems (NIPS), Montreal, QC, Canada, 7–12 December 2015.
36. Ye, M.; Ke, L.; Li, S.; Tai, Y.-W.; Tang, C.-K.; Danelljan, M.; Yu, F. Cascade-DETR: Delving into High-Quality Universal Object Detection. *arXiv* **2023**, arXiv:2307.11035.
37. Jocher, G.; Stoken, A.; Borovec, J.; Changyu, L.; Hogan, A.; Diaconu, L.; Poznanski, J.; Yu, L.; Rai, P.; Ferriday, R. ultralytics/yolov5: V3.0. *Zenodo* **2020**. [CrossRef]
38. Varghese, R.; Sambath, M. Yolov8: A novel object detection algorithm with enhanced performance and robustness. In Proceedings of the 2024 International Conference on Advances in Data Engineering and Intelligent Computing Systems (ADICS), Chennai, India, 18–19 April 2024; pp. 1–6.
39. Lei, M.; Li, S.; Wu, Y.; Hu, H.; Zhou, Y.; Zheng, X.; Ding, G.; Du, S.; Wu, Z.; Gao, Y. YOLOv13: Real-Time Object Detection with Hypergraph-Enhanced Adaptive Visual Perception. *arXiv* **2025**, arXiv:2506.17733.

plants

MDPI

Article

Comparative Leaf Proteome Analysis of Maize (*Zea mays* L.) Exposed to Combined Drought and Heat Stress

Cleopatra Pfunde [1], Charles Shelton Mutengwa [1], Graeme Bradley [2] and Nyasha Esnath Chiuta [1,*]

[1] Department of Agronomy, Faculty of Science and Agriculture, University of Fort Hare, Alice 5700, South Africa; cmutengwa@ufh.ac.za (C.S.M.)
[2] Department of Biochemistry and Microbiology, University of Fort Hare, Alice 5700, South Africa
* Correspondence: nyashachiuta@gmail.com

Abstract: This study sought to screen 45 maize (*Zea mays* L.) inbred lines for tolerance to combined drought and heat stress (CDHS) and identify the leaf proteome patterns of two inbred lines with contrasting stress response at early vegetative stage. Biomass accumulation was significantly reduced under CDHS compared to optimum conditions. Furthermore, CDHS-tolerant inbred lines exhibited significantly lower ($p < 0.05$) leaf temperatures (28.6 °C) and higher sub-stomatal CO_2 concentration (9012 mol mol^{-1}) and photosynthetic yield (0.69) under stress. The tolerant (CIM18) and susceptible (QS21) inbred lines were exposed to stress by maintaining a field capacity of 25% for 7 days and increasing the daily ambient temperature by 5 °C from 25 °C to 40 °C. Conventional two-dimensional electrophoresis analysis was used to compare leaf protein expression profiles, and significant differences ($p < 0.05$) were observed. Out of a total of 505 proteins, 114 showed significant quantitative variation. Of these, 62 proteins had a twofold upregulation in CIM18, while 52 were downregulated. Twenty upregulated proteins were selected for amino acid microsequencing, and 11 proteins were uniquely expressed in CIM18. The other nine proteins had $\geq$ twofold upregulation in CIM18 compared to QS21. The functions of the identified proteins included defence, metabolism, photosynthesis and structure.

Keywords: combined stress; climate-resilient; physiological traits; proteomics

1. Introduction

Maize (*Zea mays*) is the third most important cereal crop and plays a significant role in economic development and global food security [1]. However, its production encounters a wide array of abiotic stressors, such as combined drought and heat stress (CDHS). Maize is highly sensitive to severe CDHS at early vegetative stages, which can greatly reduce total crop stand and increase its vulnerability to pests and diseases [2]. Grzesiak et al. [3] reported a close correlation between maize seedling growth and grain yield under drought stress. Combined drought and heat stress presents a unique set of challenges, as the simultaneous occurrence of these conditions can exacerbate the negative effects of each stress on maize physiology. These stressors can cause protein dysfunction, which severely impacts maize growth and development, subsequently reducing grain yield [4]. This necessitates the development of tolerant varieties, especially, in the Sub-Saharan region, where both stressors are quite prominent early in the season.

Adaptability to drought or heat stress in maize has been attributed to various physiological traits such as photosynthetic efficiency, stomatal conductance, leaf water potential and antioxidant enzyme activity [4]. By investigating these physiological variables, researchers can gain insights into the mechanisms underlying maize's resilience to CDHS.

This knowledge is essential for developing strategies to improve maize tolerance and ensure sustainable agricultural productivity in the face of climate change.

Generally, plants respond uniquely to the occurrence and severity of abiotic stressors at different stages of crop growth, thereby increasing the complexities of breeding for climate-resilient varieties. However, different omics approaches, such as transcriptomics, metabolomics, phenomics, proteomics, etc., have significantly advanced smart breeding programs [5]. The proteomic approach is an important technique which provides great perceptions and understanding of plant responses to abiotic stressors at the protein level [6], capturing cell activity at a given time [5]. This method focuses on the actively translated portion of the genome, thus providing information missing in deoxyribonucleic acid (DNA) or messenger RNA analysis methods [7].

Proteomic analysis has been used to screen genotypes for drought and/or heat stress tolerance [6]. Comparative protein analysis has been conducted in maize [8], wheat [9], soybean [10] and several field crops exposed to drought stress [11,12]. Limited water stress can result in modification of several proteins such as aquaporins, dehydrins, heat shock proteins and late embryogenesis abundant (LEA) proteins [6]. Dehydrins are known to accumulate during late embryogenesis and on vegetative tissues of stressed (i.e., via drought, heat, salinity and cold) plants. They are therefore a group of LEA proteins that are synthesized more abundantly under stress and induce accumulation of abscisic acid (ABA). There is substantial evidence in the literature that shows how the accumulation of dehydrins confers tolerance to several abiotic stressors, including drought [13].

Heat stress influences the structure of cell membranes and physiological processes such as respiration and photosynthesis [14]. When temperatures increase, normal protein synthesis declines and stress-related proteins such as heat shock proteins (HSPs) accumulate, allowing the plant to adjust, which is necessary for conferring tolerance [15]. These proteins can be classified into five families based on their molecular masses: HSP100, HSP90, HSP70, HSP60 and small HSPs. The more abundant and diverse the HSPs, the more adapted a plant is to heat stress [16].

Proteomic analysis is increasingly becoming a biomarker of choice in marker-assisted selection (MAS) studies, where the unique protein associated with a trait is used to identify tolerant genotypes [17]. Proteomic research into maize plants subjected to individual abiotic stressors is well documented, but far fewer investigations have been conducted under combined stresses [18]. Previous proteomic studies have indicated that gene expression maybe altered, resulting in protein expression changes in plants subjected to heat and drought stress.

Over the years, a lot of technological advancement with regard to proteomic analysis, from conventional gel electrophoresis to label-free protein identification, has been observed [19]. Several proteomic technologies, such as mass spectrometry, protein arrays, affinity proteomics, ELISA and Western blot, have been developed [20]. However, the separation of expressed proteins using the convectional two-dimensional electrophoresis method is still useful. Proteomics relies on readily available genome sequence data to identify proteins of interest. Where sequences are not available, the proteins can be identified through similarity searches using homologous proteins from closely related species. Therefore, the objective of this study was to screen maize inbred lines for tolerance to CDHS and to identify proteins associated with maize tolerance at the seedling stage.

2. Results

2.1. *Effect of Combined Drought and Heat Stress on Physiological Traits of Maize and Biomass*

2.1.1. Analysis of Variance

Sub-stomatal CO_2 (C_i), leaf temperature (T_{leaf}), photosynthetic yield (PSII) and shoot and root dry weight were highly significantly different ($p < 0.001$) among inbred lines exposed to CDHS (Table A1).

2.1.2. Sub-Stomatal CO_2

Combined drought and heat stress exhibited a significant ($p < 0.001$) increase in CO_2 uptake compared to optimum conditions. Inbred line CIM 20 recorded the highest sub-stomatal CO_2 of 9012 mol mol^{-1} under CDHS (Table 1). However, QS21 recorded significantly lower sub-stomatal CO_2 (531 mol mol^{-1}) and was part of the bottom five lines that showed great sensitivity to combined stress (Table 1). On the other hand, CIM 18 showed moderate tolerance by recording a sub-stomatal CO_2 of 3403 mol mol^{-1}.

Table 1. Mean physiological changes and biomass of the top ten and bottom ten genotypes under combined drought and heat stress.

Rank	PSII		Ci (mol mol^{-1})		T Leaf (°C)		SDW (g)		RDW (g)	
1	QS17	0.69 ^{a}	CIM20	9012.00 ^{a}	CIM15	35.10 ^{a}	QS22	0.17 ^{a}	CIM12	0.09 ^{a}
2	QS1	0.69 ^{ab}	QS23	9012.00 ^{a}	QS25	34.60 ^{a}	CIM12	0.14 ^{ab}	QS22	0.08 ^{ab}
3	QS30	0.68 $^{a–c}$	CIM9	5506.00 ^{b}	QS1	34.35 ^{ab}	CIM18	0.14 $^{a–c}$	CIM18	0.08 $^{a–c}$
4	QS16	0.67 $^{a–c}$	CIM13 ^{b}	5506.00	QS32	34.30 ^{ab}	CIM20	0.14 $^{a–c}$	CIM10	0.08 $^{a–c}$
5	CIM18	0.67 $^{a–c}$	CIM21	3935.00 ^{c}	QS18	33.95 ^{ab}	QS8	0.14 $^{a–c}$	QS8	0.07 $^{a–d}$
6	CIM14	0.67 $^{a–c}$	CIM5	3935.00 ^{c}	CIM10	33.80 ^{ab}	QS26	0.14 $^{a–d}$	QS26	0.07 $^{a–d}$
7	CIM11	0.67 $^{a–c}$	CIM6	3935.00 ^{c}	QS29	33.80 ^{ab}	QS18	0.13 $^{a–e}$	QS18	0.07 $^{b–e}$
8	CIM12	0.67 $^{a–c}$	QS3	3935.00 ^{c}	CIM16	33.55 $^{a–c}$	CIM10	0.13 $^{b–f}$	CIM7	0.06 $^{b–f}$
9	CIM1	0.66 $^{a–c}$	QS5	3789.00 ^{d}	CIM6	33.50 $^{a–c}$	QS6	0.12 $^{b–f}$	QS5	0.06 $^{c–g}$
10	QS7	0.66 $^{a–c}$	QS20	3591.00 ^{e}	CIM21	33.40 $^{a–d}$	QS7	0.12 $^{b–f}$	QS6	0.06 $^{c–g}$
			Bottom							
1	CIM2	0.60 $^{a–e}$	CIM16	827.00 ^{w}	CIM19	31.10 $^{b–j}$	QS16	0.09 $^{e–i}$	QS28	0.04 $^{f–h}$
2	QS19	0.60 $^{a–e}$	QS15	827.000 ^{w}	QS23	31.05 $^{b–j}$	QS15	0.09 $^{e–i}$	CIM2	0.04 $^{f–h}$
3	QS28	0.57 $^{a–e}$	QS16	772.000 ^{x}	QS21	30.30 $^{c–j}$	CIM15	0.09 $^{e–i}$	CIM11	0.04 $^{f–h}$
4	CIM16	0.53 $^{a–e}$	CIM17	544.000 ^{y}	QS8	30.10 $^{d–j}$	CIM17	0.09 $^{f–i}$	CIM13	0.04 $^{f–h}$
5	CIM21	0.52 $^{a–e}$	QS21	531.000 ^{Z}	CIM7	29.80 $^{e–j}$	CIM21	0.09 $^{f–i}$	CIM16	0.04 $^{f–h}$
6	QS4	0.49 $^{b–e}$	QS27	531.000 ^{z}	QS27	29.35 $^{f–j}$	QS10	0.09 $^{f–i}$	QS1	0.04 $^{f–h}$
7	CIM13	0.49 $^{c–e}$	CIM1	438.00 ^{za}	CIM20	29.00 $^{g–j}$	QS23	0.09 $^{f–i}$	QS 10	0.04 $^{f–h}$
8	CIM19	0.43 ^{de}	CIM2	361.00 ^{zb}	QS17	28.70 $^{h-j}$	CIM19	0.08 $^{g–i}$	QS15	0.04 ^{gh}
9	CIM5	0.40 ^{e}	QS4	361.00 ^{zb}	CIM18	28.60 ^{ij}	CIM1	0.08 ^{hi}	CIM19	0.04 ^{gh}
10	QS5	0.08 ^{f}	QS6	−793.00 ^{zc}	CIM12	28.30 ^{ij}	QS21	0.07 ^{i}	CIM1	0.03 ^{h}

PSII —ΦPSII = (Fm′ − Fs)/Fm′; Ci—sub-stomatal CO_2; Tleaf—leaf temperature; SDW—shoot dry weight; RDW—root dry weight. Means followed by the same letter in a column were not significantly different at ($p \leq 0.05$) according to Tukey's HSD test.

2.1.3. Leaf Temperature

Leaf temperature was significantly ($p \leq 0.05$) elevated by 3–8 °C in plants exposed to CDHS compared to the control. Stressed plants continued to show higher leaf temperatures throughout the experiment. Inbred lines CIM12 and CIM18 had the lowest leaf temperatures under stresses of 28.3 °C and 28.6 °C, respectively (Table 2). Contrariwise, sensitive lines recorded leaf temperatures above 35 °C.

Table 2. Assessment of combined heat and drought stress based on tolerance indices.

INBRED LINE	STI	DRI	K2STI	Inbred Line	STI	DRI	K2STI
CIM1	0.16	0.15	0.09	QS14	0.26	0.17	0.21
CIM18	**0.32**	0.42	**0.58**	QS15	0.23	0.11	0.13
CIM11	0.25	0.44	0.40	QS16	0.18	0.18	0.13
CIM12	0.22	**0.61**	0.40	QS17	0.26	0.28	0.31
CIM13	0.16	0.27	0.14	QS18	0.30	0.30	0.41
CIM14	0.20	0.22	0.17	QS19	0.28	0.26	0.33
CIM15	0.16	0.15	0.09	QS20	0.25	0.18	0.21
CIM16	0.19	0.17	0.13	QS21	0.21	0.28	0.21
CIM17	0.21	0.12	0.12	QS22	**0.43**	**0.54**	**1.07**
CIM10	0.16	0.15	0.09	QS23	0.17	0.14	0.10
CIM2	0.12	0.14	0.05	QS25	0.25	0.23	0.25
CIM20	0.18	0.25	0.15	QS26	0.29	0.32	0.39
CIM21	0.21	**0.54**	0.33	QS27	0.22	0.21	0.18
CIM19	0.14	0.17	0.08	QS28	0.20	0.16	0.14
CIM3	0.21	0.28	0.21	QS29	0.22	0.20	0.19
CIM4	0.22	0.21	0.18	QS30	0.31	0.19	0.31
CIM5	0.28	0.26	0.33	QS32	0.21	0.16	0.15
CIM6	0.26	0.22	0.26	QS4	0.22	0.15	0.16
CIM7	0.30	0.24	0.35	QS5	0.32	0.23	0.38
CIM8	0.27	0.21	0.27	QS6	**0.35**	0.26	**0.47**
CIM9	0.24	0.19	0.20	QS7	0.25	0.29	0.29
QS10	0.20	0.16	0.14	QS8	0.26	0.43	0.41
QS1	0.26	0.22	0.26				

The highest data for each parameter is exhibited in bold. STI—Stress tolerance index; DRI—drought resistance index; K_2 STI—modified stress tolerance index.

2.1.4. Photosynthetic Yield (PSII)

Efficiency of photosynthetic yield was significantly ($p < 0.001$) higher in control plants compared to those under stressed conditions throughout the experiment. Photosynthetic yield efficiency was 17–29% less under CDHS than in control plants. Inbred lines QS17 and QS1 had the highest photosynthetic yield during stress, closely followed by QS30, QS16 and CIM 18 (Table 2). After the recovery period, differences in PSII efficiency were non-significant ($p > 0.05$).

2.1.5. Shoot and Root Dry Weight

Shoot and root dry weight were significantly reduced by CDHS. Inbred lines CIM 12, QS22 and CIM 18 were the top three best performers for shoot and root dry weight under CDHS. Inbred line QS 21 recoded the lowest dry shoot weight (0.07 g) and moderated dry root weight (0.05 g) under stress. Generally, shoots were more severely affected by stress compared to roots.

Although the primary focus was on stress-induced changes, the inclusion of a 3-day recovery period allowed for assessment of resilience. Notably, PSII efficiency differences between CIM18 and QS21 were no longer significant post-recovery, indicating partial restoration of photosynthetic function. The higher final biomass and upregulation of stress-responsive proteins in CIM18 suggest a more robust recovery mechanism compared to QS21. These findings highlight the importance of recovery capacity as a component of stress tolerance.

2.1.6. Stress Tolerance Indices

Shoot dry weight was used to calculate different indices, namely stress tolerance index (STI), drought resistance index (DRI) and modified stress tolerance index (K_2STI). Based on the results, QS22, QS6 and CIM18 exhibited CDHS tolerance, as indicated by the higher index values recorded (Table 2).

2.1.7. Quantitative Comparative Analysis of Protein Responses

Comparative proteomic analysis was used to investigate the change in protein profiles in leaves of tolerant (CIM18) and susceptible (QS21) maize inbred lines after exposure to CDHS. Thus, this study focused on comparative assessment of proteins differentially expressed by these two inbred lines under CDHS. Approximately 505 spots were reproducibly detected in each Sypro Ruby-stained gel. The partial least squares analysis revealed that 185 protein spots were significantly different ($p < 0.05$) between CIM18 and QS21. Of these, 114 proteins exhibited a twofold differential expression (increase/decrease) between the tolerant and susceptible inbred lines. A total of 62 of these were either upregulated or newly induced in response to CDHS, while 52 were downregulated in CIM18. Of the 114 proteins, QS21 had 52 proteins upregulated and 62 downregulated. Twenty-four protein spots that showed a twofold increase or more in CIM18 and exhibited a coefficient of variation between 0% and 2% are shown in Figure 1.

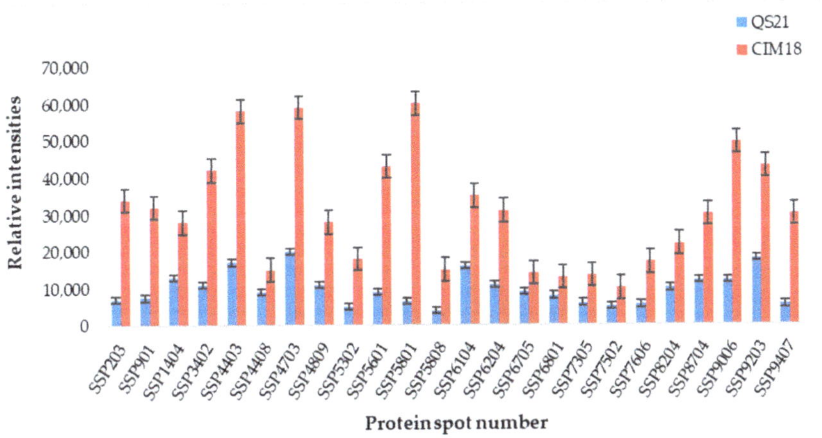

Figure 1. Twenty-four protein spots which significantly increased twofold or more in the tolerant (CIM18) compared to susceptible (QS21) inbred line under combined drought and heat stress treatment. Data shown are means ± SD of three biological replicates.

Examples of gel pictures showing protein upregulation are shown in Figure A1, whereas the functional categories of proteins uniquely upregulated in the CDHS-tolerant inbred line CIM18 are shown in Figure 2.

The flow chart provides a clear overview of how the proteins were grouped into broader categories: photosynthesis, metabolism, stress response, protein biosynthesis and cytoskeleton organization. Eleven protein spots were unique only to the tolerant inbred line (CIM18), as shown in Figure A2. The percentage of proteins uniquely upregulated in the tolerant CIM18 inbred line is shown in Figure 3.

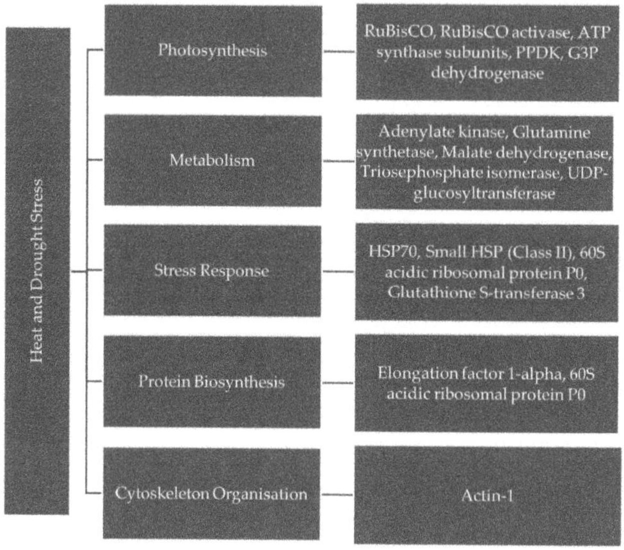

Figure 2. Functional categories of proteins uniquely upregulated in the combined drought and heat stress-tolerant inbred line, CIM18.

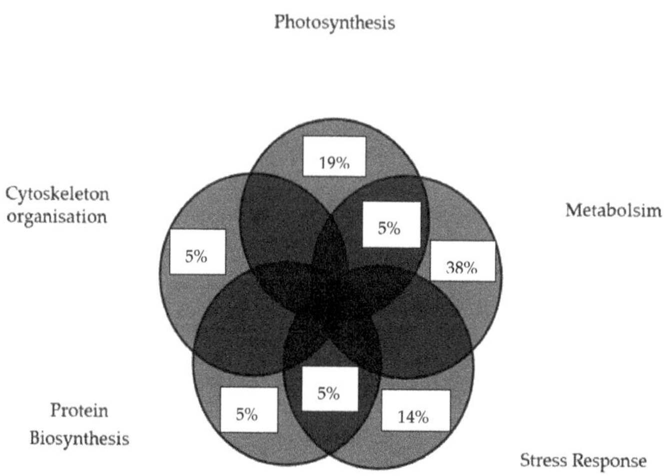

Figure 3. Percentage of proteins uniquely upregulated in the combined drought and heat stress-tolerant inbred line, CIM18, according to functional category.

The identified spots were classified according to their putative functions. The differentially expressed proteins and their individual functions are detailed in Table 3.

Table 3. Identification of differentially responsive proteins in maize leaves subjected to combined drought and heat stress.

Spot #	Protein Names	Accession	% Coverage	Subcellular Location	Function	GO Annotation
4606	Actin-1	sp I P02582 I ACT1_MAIZE	9.33	Nucleus (GO:0005634) Cytoskeleton (GO:0005856)	Cytoskeleton organization	GO:0007010
8406	Adenylate kinase, chloroplastic	sp I P43188 I KADC_MAIZE	58.11	Plastid; chloroplast (GO:0009507)	Metabolism signalling	GO:0006139 GO:0046940
7403	Glyceraldehyde-3-phosphate dehydrogenase A, chloroplastic	sp I P09315 I G3PA_MAIZE	8.933	Plastid; chloroplast (GO:0009507)	Metabolism, photosynthesis	GO:0006006, GO:0006006 GO:0009416, GO:0019253
4607	DIMBOA UDP-glucosyltransferase BX9	sp I B4G072 I BX9_MAIZE	36.80	Cytoplasm (GO:0005737)	Metabolism	GO:0008194, GO:0016740 GO:0016757, GO:0047254 GO:0080043, GO:0080044
8509	Glyceraldehyde-3-phosphate dehydrogenase 2, cytosolic	sp I Q09054 I G3PC2_MAIZE	48.66	Cytoplasm (GO:0005737)	Metabolism	GO:0006006, GO:0006096
4605	Glutamine synthetase root isozyme 3	sp I P38561 I GLNA3_MAIZE	22.47	Cytoplasm (GO:0005737)	Metabolism	GO:0006542
5605	Malate dehydrogenase [NADP], chloroplastic	sp I P15719 I MDHP_MAIZE	18.98	Plastid; chloroplast (GO:0009507)	Metabolism	GO:0006099, GO:0006107 GO:0006108, GO:0019674
0405	Glutamine synthetase, chloroplastic	sp I P25462 I GLNAC_MAIZE	35.70	Plastid; chloroplast (GO:0009507)	Metabolism	GO:0006542
1606	ATP synthase subunit beta, mitochondrial	sp I P19023 I ATPBM_MAIZE	1.989	Mitochondrion; mitochondrion inner membrane (GO:0005743)	Metabolism; ATP metabolism	GO:0046034, GO:0006754 GO:0015986, GO:0042776
5404	Triosephosphate isomerase, cytosolic	sp I P12863 I TPIS_MAIZE	22.13	Cytoplasm (GO:0005737)	Metabolism; carbohydrate metabolism	GO:0006094, GO:0006096 GO:0016052
8611	Ribulose bisphosphate carboxylase large chain	sp I P00874 I RBL_MAIZE	49.16	Plastid; chloroplast (GO:0009507)	Photosynthesis	GO:0009853, GO:0015977 GO:0015979, GO:0019253
5609	Ribulose bisphosphate carboxylase/oxygenase activase, chloroplastic	sp I Q9ZT00 I RCA_MAIZE	19.63	Chloroplast stroma (GO:0009570)	Photosynthesis	GO:0046863, GO:0016887 GO:0005524
4807	Pyruvate, phosphate dikinase 1, chloroplastic	sp I P11155 I PPDK1_MAIZE	28.83	Plastid; chloroplast (GO:0009507) Cytoplasm (GO:0005737)	Photosynthesis	GO:0015979
8503	ATP synthase subunit gamma, chloroplastic	sp I P0C1M0 I ATPG_MAIZE	42.34	Plastid; chloroplast thylakoid membrane (GO:0009535)	Photosynthesis Proton transmembrane transport	GO:0006754, GO:1902600
3503	60S acidic ribosomal protein P0	sp I O24573 I RLA0_MAIZE	19.12	Ribosome (GO:0005840) Cytosolic large ribosomal subunit (GO:0022625)	Protein biosynthesis Stress response to anoxia	GO:0002181 GO:0034059
0406	Elongation factor 1-alpha	sp I Q41803 I EF1A_MAIZE	6.711	Cytoplasm (GO:0005737)	Protein biosynthesis	GO:0006412 GO:0006414
5205	17.8 kDa class II heat shock protein	sp I P24632 I HSP22_MAIZE	37.20	Cytoplasm (GO:0005737)	Stress response, protein folding Heat response, salt stress response Hydrogen peroxide response Protein oligomerization	GO:0006950, GO:0006457 GO:0009408, GO:0009651 GO:0042542, GO:0051259
4804	Heat shock 70 kDa protein	sp I P11143 I HSP70_MAIZE	19.22	Cytoplasm (GO:0005737)	Stress response to heat Protein folding	GO:0009408 GO:0042026
7304	Glutathione S-transferase 3	sp I P04907 I GSTF3_MAIZE	16.67	Cytoplasm (GO:0005737)	Stress response to herbicide	GO:0009635

3. Discussion

Plants exhibit stress tolerance or avoidance through acclimation and adaptation mechanisms. This study identified QPM inbred lines that exhibited tolerance to CDHS using physiological traits such as gaseous exchange, photosynthetic yield and leaf temperature. A wide genetic variation existed among the evaluated inbred lines. Similar screening and characterization studies are useful in preliminary elimination of susceptible lines in breeding programs involving several genotypes.

Generally, drought reduces internal CO_2 due to stomata closure, whereas heat stress increases sub-stomatal CO_2 concentration due to reduced efficacy of ribulose bisphosphate carboxylase/oxygenase (RUBISCO) activase caused by elevated leaf temperatures [21]. As such, CDHS can cause complexities in overall sub-stomatal CO_2 concentration [22]. In this study, CIM 18 exhibited higher sub-stomatal CO_2 concentration relative to QS21. Additionally, CIM 18 recorded lower leaf temperature and higher photosynthetic yield. Similarly, the ability to balance between conserving water by closing the stomata or opening the stomata to allow transpirational cooling and CO_2 influx was observed on CDHS-tolerant broadleaf evergreen shrubs [22]. Undoubtedly, the high sub-stomatal CO_2 concentration and low leaf temperature in CIM 18 positively influenced its photosynthetic yield under stress. Conversely, QS21 exhibited stress susceptibility by recording high leaf temperatures and low CO_2 flow and photosynthetic rate under stress.

Consistent with other studies [23], shoot and root dry weight accumulation was inhibited by stress conditions, with the shoots being more sensitive than the roots. As such, dry shoot weight was used to calculate the different stress indices. High values of STI, DRI and K_2STI indicated tolerant genotypes. Among the inbred lines evaluated, CIM18 showed moderate tolerance and ranked third, fifth and second in STI, DRI and K_2STI, respectively. However, QS21 showed great susceptibility to CDHS and was ranked in the bottom 10 for all the indices calculated. Additionally, CIM 18 showed tolerance to heat stress alone in previous laboratory and greenhouse studies [24], and thus could possess genes for CDHS tolerance.

It was hypothesised that comparative analysis of tolerant and susceptible genotypes would result in identification of proteins found under normal growth conditions, those that respond to stress, and those involved in tolerance response. Exposure of CIM18 and QS21 to CDHS in the current study resulted in identification of some proteins with well-known functional roles in plants exposed to abiotic stresses. This allowed us to infer their potential contribution towards the tolerance of CIM118 to CDHS. Other studies have also focused on comparative assessment of transcriptomes differentially expressed by genotypes with contrasting responses to abiotic and biotic stresses. For example, Subramani et al. [25] conducted a comparative analysis of untargeted metabolomics in tolerant and sensitive genotypes of *P. vulgaris* seeds exposed to terminal drought stress. They conducted different pairwise comparisons of drought-tolerant and -sensitive genotypes that were only subjected to drought stress, without exposure to optimum conditions. In their case, the comparisons involved three tolerant and three susceptible genotypes. The comparisons led to successful identification of metabolites associated with drought tolerance. Another study revealed that sweet potato genotypes with contrasting responses to salt stress had distinct differences at the transcription level and translation level even without salt stress [26]. It was reported that proteins that were differentially more abundant in the tolerant type, some of which were absent from the susceptible genotype, were contributing to salt tolerance. Tomato genotypes with contrasting responses to late blight were also used for comparative analysis of their constitutive proteome, without exposing them to stress [27]. They found two defence proteins that were more abundant in the resistant genotype, and these were assumed to be responsible for resistance against late blight.

The proteins identified were involved in numerous biological pathways, which directly and indirectly impact plant protection. The number of proteins micro-sequenced was limited only to those uniquely expressed and upregulated in the tolerant line due to the cost of protein spot identification. Some of the omitted proteins could have likely contributed to tolerance observed in CIM18. Approximately 8.8% more spots were upregulated in CIM18 in comparison to QS21. Subcellular localization revealed that the upregulated proteins were synthesized in the chloroplast and cytoplasm. Therefore, most of these proteins were essential for photosynthesis and energy-related activities, suggesting adaptation to the combined stresses. Similar observations were made in date palm [28] and soybean [29] exposed to CDHS. The identification of proteins and their subcellular locations helps in understanding their physiological function.

A decline in protein abundance reflected cellular damage caused by the two stressors. This included Ribulose- 1,5- bisphosphate carboxylase/oxygenase (Rubisco), Rubisco activase enzymes and other key enzymes of the Calvin cycle, such as phosphoglycerate kinase and fructose–bisphosphate aldolase 1, which catalyses carboxylation. The downregulation of these enzymes is likely responsible for the reduced growth and development observed in QS21 compared to CIM18.

3.1. *Photosynthesis- and Metabolism-Related Proteins*

Rubisco (spot 8611) was one of the proteins associated with photosynthesis that was identified in this study. The increased levels of rubisco in carbon fixation could have been a result of increased energy needs under CDHS. Ribulose biphosphate carboxylase/oxygenase activase is a molecular chaperone responsible for switching Rubisco from an inactive form to an active form. It has been proposed that the reason for the upregulation of these photosynthesis-related proteins is to alleviate damage to Rubisco which is caused by the abiotic stressors [30].

Subunit components of the adenosine triphosphate (ATP) synthase chloroplast protein complex that are involved in ATP synthesis, such as the alpha, gamma and beta subunits, were upregulated in this study. ATP synthase is responsible for the production of ATP from ADP. Adenosine triphosphate is used for various energy-demanding activities during stress, such as protein degradation and biosynthesis, as evidenced by the upregulation of elongation factor 1 alpha, 60S acidic ribosomal protein and HSP70. ATP synthase likely contributed to the extended survival time of plant cells that had become ATP-depleted due to stress [7]. This showed that there was likely activation of ATP-generating pathways under combined stress to meet the increasing demand of ATP to maintain homeostasis. Higher levels of ATP synthase were reported by Loka et al. [31] in cotton subjected to CDHS. The abundance of these proteins and the activities of the Calvin cycle enzymes in CIM18 could have prevented a major reduction in the level of photosynthesis [29].

Photosynthesis depends on Rubisco and the regeneration of RuBP. Results indicated that the Rubisco large chain was upregulated in CIM18 compared to QS21, and this resulted in maintenance of photosynthesis and growth. Although Rubisco occurs abundantly in plant leaf tissue, the combined stress imposed caused its quantity and activity to decrease drastically in susceptible QS21 causing its detection to be difficult. Other reports have shown similar findings under drought or CDHS on non-tolerant plants [30]. The findings support the physiological analysis of CIM18 in another study, where it was ranked among the top 10 inbred lines exhibiting high root to shoot ratio under CDHS [24].

Proteins related to carbohydrate metabolism showed altered expression patterns induced by CDHS. A review by Xu et al. [18] denoted that the concentration of most soluble sugars increased under CDHS and could be used to explain variable tolerance among cultivars. The protein changes related to carbohydrate metabolism were necessary for the adaptation of seedlings to the combination of the two stresses. Carbohydrate metabolism is the main bio-molecular metabolism, and the substrates involved provide energy required by the plant [18,32]. Enzymes related to energy metabolism responded to CDHS in CIM18. Malate dehydrogenase, a protein enzyme involved in carbohydrate synthesis, was upregulated. The abundance of other metabolism-related proteins such as uridine diphosphate glucosyltransferase and glutamine synthetase increased, suggesting a metabolic change in leaves of CIM18 plants under combined stress. Glutathione synthetase showed increased abundance following combined stress in CIM18. These enzymes are important for proline biosynthesis and regulation of bioactivities; their accumulation indicates tolerance to abiotic stressors [8].

It is known that pyruvate phosphate dikinase is a key enzyme in gluconeogenesis and photosynthesis. It reverses the reaction performed by pyruvate kinase in glycolysis and its expression has been associated with tolerance to several abiotic and biotic stressors [33]. Gluconeogenesis requires a lot of energy, which is supplied by the ATP-generating pathways. During exposure to the combined stresses, we observed that there was an increase in the expression of the elongation factor 1-alpha protein, also known as elongation factor thermo-unstable (EF-Tu) protein. The elongation factor thermo-unstable protein was identified as a potential biomarker of heat stress tolerance in soybeans [30]. This protein is an essential component of protein synthesis, and its overexpression greatly improves plant heat stress responses [34].

3.2. Heat-Stress-Responsive Proteins

Two heat shock protein families, HSP70 and small HSP 17.8, were identified as stress-related proteins [35]. Heat shock proteins are induced under various stress conditions, but particularly when temperatures exceed 35 °C (temperatures often experienced in the field). In this study, HSPs mediated tolerance to CDHS. These proteins played a significant role in conferring tolerance in CIM18. Several studies have alluded that enhancement of thermo-tolerance is achieved through synthesis of heat shock proteins [14,15]. The findings suggested that drought and heat tolerance was correlated with the accumulation of small heat shock proteins (sHSPs). Small HSP 17.8, uniquely identified in CIM18 in this study, has also been reported by Klein et al. [36] in maize and cardamom [37] subjected to heat stress. The role of HSPs is to stabilize the heat-stressed membranes and proteins while encouraging protein refolding under stress [15]. The tolerant CIM18 inbred line showed increased abundance of HSP70. The function of HSP70 includes protein refolding to maintain cellular homeostasis [38]. Higher expressions of HSP70 have been established in soybean and tomato plants subjected to CDHS [39,40]. Heat shock protein 70 has been widely used as a biomarker under different stressors [41]. However, earlier studies associated HSP with response to heat stress alone instead of other abiotic and biotic stresses.

3.3. Antioxidant Proteins

Combined drought and heat stress-induced oxidative stress results in the accumulation of reactive oxygen species (ROS) [38,42]. To scavenge and eliminate these, plants express antioxidant enzymes such as glutathione S transferase (GST). In the current maize study, GST (spot 5404) was upregulated in the tolerant CIM18 line compared with QS21. The accumulation of these antioxidant enzymes constituted part of the detoxification mechanism for excess ROS during oxidative stress. Glutathione S transferase is involved in the glutathione–ascorbate cycle, which is a scavenging system activated in maize to alleviate oxidative stress and enhance drought tolerance. The expression of GST is enhanced by abiotic and biotic stresses [32,43]. The overexpression of protein kinases such as adenylate kinase was likely responsible for the reduction in the accumulation of ROS in the leaves of CIM18.

These results corroborate with previous reports, where defence-related proteins such as antioxidant proteins were upregulated under drought stress [43,44]. A study investigating proteomic responses of maize inbred lines with contrasting responses to drought stress also revealed that the tolerant genotype experienced enhanced levels of ROS scavenging enzymes, leading to a higher ROS scavenging ability than in the susceptible genotype [45].

3.4. Structural Proteins

The drought stress treatment in this study likely induced actin 1. Actin filaments constitute the cytoskeleton of the plant cell and are responsible for the maintenance of cellular architecture and orientation of organelles [46]. The filaments may also act as osmo-sensors and target potassium ion channels in guard cells for osmoregulation under drought stress. The abundance of this protein in CIM18 could be interpreted as adaptation to drought stress. This corroborates with a report by [47], who reported drought tolerance in barley expressing actin 1.

The potential use of some of the above proteins as biomarkers may be hindered for a few reasons. Before a biomarker can be successfully applied, it must be tested against different genotypes and must be specific, reproducible and sensitive. However, as researchers continue to pursue proteins as biomarkers, they have discovered new techniques to conduct proteomic analysis that are able to overcome challenges posed by the 2-DE gel approach [48].

4. Materials and Methods

4.1. Plant Materials

Forty-five white quality protein maize inbred lines (Table A2) obtained from the Quality Seed company (Pietermaritzburg, South Africa) and CIMMYT (Harare, Zimbabwe) were assessed for tolerance to CDHS in controlled growth chambers in the Biochemistry and Microbiology Laboratory at the University of Fort Hare, Alice, South Africa. Two inbred lines (QS21 and CIM18) showing contrasting tolerance to CDHS were then used in the proteomic study.

4.2. Experimental Design

Forty-five inbred lines were screened for tolerance to CDHS conditions prior to the proteomic experiment. These inbred lines were laid out in a 9×5 randomized incomplete block design, with two replications. Each inbred line was replicated three times in each block, and three biological replications (cycles) were performed for each stress condition. The control (no stress imposed) and CDHS experiment lines were grown in separate growth chambers.

In the second experiment, the two selected inbred lines with constating behaviour under CDHS were laid out in a completely randomised design with three replicates and exposed to CDHS treatment. Three pots represented each inbred line per replication, and each experiment was replicated three times over two cycles. In both experiments, plants were grown from seed in pots filled with hygromix growing media (commercial potting mix). Seedlings were thinned to one plant/pot three days after emergence.

4.3. Drought and Heat Treatment

The growth chambers were programmed to provide a 14 h/10 h day/night diurnal cycle at an air CO_2 concentration of 400 μmol mol−1 and 40% relative humidity (RH) [49]. Plants were exposed to drought stress at seven days after emergence by maintaining a field capacity (FC) of approximately 25% for seven days. An SM300 soil moisture meter (Delta T Devices, Cambridge, UK) was used to measure soil moisture content, and plants were irrigated with hygrofert when FC dropped below 25%. Heat stress was imposed on seedlings at the three-leaf stage. Temperature was increased gradually from 25 °C/20 °C (day/night) in 5 °C increments per day to a max of 40 °C/35 °C. The plants were maintained at maximum temperatures for one day. Plants in the high-temperature-regime growth chambers were returned to the control growth chamber (25 °C) for three days to recover [50].

In the control growth chamber, plants were maintained at 75% FC and irrigated once daily with a complete nutrient solution (hygrofert). The growth chambers were programmed to provide a 14 h/10 h day/night diurnal cycle. Air temperature was 25 °C/22 °C, air CO_2 concentration was maintained at 400 μmol mol^{-1} and RH was 60% for the control and 40% for CDHS. Both experiments were terminated 21 days after initiation [8,18,31].

4.4. Data Collection

4.4.1. Physiological Traits

Single-leaf photosynthetic and gaseous exchange measurements were performed with an Integrated Fluorometer and Photosynthesis system (ADC BioScientific Ltd, Hertfordshire, UK). Sub-stomatal CO_2 (mol mol^{-1}), photosynthetic active radiation (Watts m^{-2}) and photochemical efficiency of photosystem II ($\Phi PSII = (Fm' - Fs)/Fm'$) measurements were taken in the middle of the third fully developed leaf before, during and after the stress period. The effective quantum yield of PSII was determined on light-adapted leaves during steady-state photosynthesis.

4.4.2. Plant Growth

Shoots and roots were carefully separated, and the latter were thoroughly washed to remove dirt and then oven-dried at 60 °C until a constant weight was obtained. Dry shoot data were used to calculate stress tolerance indices.

4.5. Protein Extraction and Quantification

Leaf samples from three plants per genotype and treatment were pooled and ground in liquid nitrogen. Proteins were extracted using the ReadyPrep Protein Extraction Kit (Bio-Rad Laboratories Inc, Sandton, South Africa), followed by sonication and centrifugation at 16,000× g for 30 min at 20 °C. The supernatant was collected and standardized to 100 μg using the DC Protein Assay Kit (Bio-Rad), with concentrations determined from a bovine serum albumin (BSA) standard curve.

4.6. Protein Clean-Up and Two-Dimensional Electrophoresis

Protein samples were cleaned using the ReadyPrep 2-D Cleanup Kit (Bio-Rad Laboratories Inc, Sandton, South Africa), precipitated, washed and resuspended in rehydration buffer. Proteins (100 μg) were loaded onto 7 cm IPG strips (pH 4–7) and rehydrated overnight. Isoelectric focusing was followed by SDS-PAGE on 12.5% gels. Gels were stained with SYPRO Ruby and visualized using a Uvitec gel documentation system.

4.7. Gel Processing and Peptide Analysis

Excised protein spots were destained, dehydrated and subjected to reduction and alkylation using DTT and iodoacetamide. Trypsin digestion was performed overnight at 37 °C. Peptides were extracted using acetonitrile/formic acid, vacuum-dried and cleaned using stage tips for MALDI-TOF analysis.

5. Data Analysis

The data presented represents averaged results of three cycles of the 45 inbred lines for every measured parameter. Bartlett's test was performed to test the homogeneity of error variances for each of the three cycles. The test was significant, and data were combined. Data were found to be normally distributed using Shapiro–Wilk's test and then subjected to a two-way ANOVA when the conditions and the assumptions of repeated measures were met. The conditions imply sphericity, which means that the variances of the repeated measures are equal and the correlations among the repeated measures are equal. The Tukey HSD test was used to perform post hoc ANOVA comparisons using JMP version 16. Dry shoot weights were used to calculate three tolerance indices using the following formulae.

$$\text{Stress Tolerance Index (STI)} = \frac{Y_P \times Y_s}{Y_P 2} \tag{1}$$

where Y_s and Y_P represent shoot dry weight under stress and non-stress conditions, [51].

$$\text{Modified Stress Tolerance Index (MSTI)} = \text{K2STI}, = \left(Y_S{}^2\right) / \left(\overline{Y}_S{}^2\right) \tag{2}$$

where Y_s and $\overline{Y}_S$ represent shoot dry weight under stress and mean yield under stress, respectively [51].

$$\text{Drought Resistance Index (DRI)} = \frac{Y_S}{Y_P} / \frac{\overline{Y}_S}{\overline{Y}_P} \tag{3}$$

where Y_s and Y_P represent shoot dry weight under stress and non-stress conditions, respectively. Also, $\overline{Y}_S$ and $\overline{Y}_P$ are the mean yield under stress and non-stress conditions, respectively [51].

Protein spot detection and matching, normalization, quantification and analysis were performed using PDQUEST software 8.0.1 (Bio-Rad). Analytical tools and algorithms in PDQUEST compared the 2-D gels to identify any significant changes in protein expression between tolerant (CIM18) and susceptible (QS21) inbred lines in response to drought and heat stress. Analytical tools in PDQuest helped to identify protein spots of interest. Advanced algorithms in PDQuest identified and matched protein patterns. The 2-D gels for CIM18 and QS21 were selected for comparison and grouped according to treatment. Preset methods for matching and normalization were used. Sophisticated algorithms automated detection and spot matching. The SYPRO Ruby filter provided auto-recognition and removal of background speckles. A spot that occurred in gels of both the tolerant and susceptible genotypes at roughly the same intensity and was identified as being neither up or downregulated by the software was selected and used as a standard for the software's assessment of whether a protein spot was up- or downregulated. This reference standard was used as the optional control since we negated the need for comparison against the susceptible non-stressed plants (control). Differences in protein abundance between individual gels in the two varieties were validated by Student's t-test analysis at $p < 0.05$. Amino acid sequences for the uniquely identified proteins in the tolerant inbred line were compared with sequences in the database by using a Uniprot-*Zea mays* database. The molecular functions of the identified proteins were classified according to their biological functions using Gene Ontology (GO) annotation from the Blast 2GO software, version 4.

6. Conclusions and Recommendations

The results presented here indicate that CDHS inhibited plant growth and biomass accumulation due to an increase in leaf temperatures and reduced sub-stomatal CO_2 concentration and photosynthetic yield. Inbred lines showing contrasting responses to CDHS were further evaluated by conducting proteomic studies. The tolerant inbred line

(CIM18) responded to CDHS by modulating the expression of proteins that have previously been associated with stress response. The upregulation of antioxidant enzymes enhanced the stress defence response of CIM18, in addition to accelerated biosynthesis of proteins such as HSPs and carbohydrate-metabolism-related proteins, all of which might have conferred tolerance to CDHS. The preliminary findings could contribute to understanding the molecular mechanisms of CDHS tolerance in maize. The differentially expressed proteins such as EF-1 alpha and HSP70 can be validated using the ELISA and Western blotting method together with quantitative real-time polymerase chain reaction. It will also be important to find out if the same stress-responsive proteins will be produced at other stages of growth and under field conditions. It is proposed that this proteomics-based knowledge be further investigated before being directly used in the improvement of CDHS tolerance in maize through technologies such as genome editing.

Author Contributions: C.P. contributed through generation of a detailed methodology, investigation, data curation, formal analysis, result interpretation, writing, editing and review of the first and final drafts. C.S.M. contributed through conceptualization, acquisition of funding for the project, supervision, validation, project management, reviewing and editing of the first and final drafts. G.B. contributed through acquisition of resources and close supervision of C.P. as she conducted this work. N.E.C. contributed through data collection, writing, editing and review of the first and final drafts. All authors have read and agreed to the published version of the manuscript.

Funding: The Research and Technology Fund (grant no. 98706) of the National Research Foundation (NRF) of South Africa is acknowledged for funding the project.

Data Availability Statement: The data presented in this study are available upon reasonable request from the corresponding author. The data are not publicly available due to [technical/time limitations insert reason here].

Acknowledgments: Quality Seed (PTY) LTD and CIMMYT-Zimbabwe are acknowledged for supplying the Quality Protein Maize germplasm used in this study.

Conflicts of Interest: The authors declare no conflicts of interest.

Abbreviations

The following abbreviations are used in this manuscript:

CDHS	Combined drought and heat stress
CIM	CIMMTY
QS	Quality Seed
CO_2	Carbon dioxide
ATP	Adenosine triphosphate
Rubisco	Ribulose- 1,5- bisphosphate carboxylase/oxygenase
HSP	Heat shock protein
GST	Glutathione S transferase
ROS	Reactive oxygen species
2-D	Two-dimensional

Appendix A

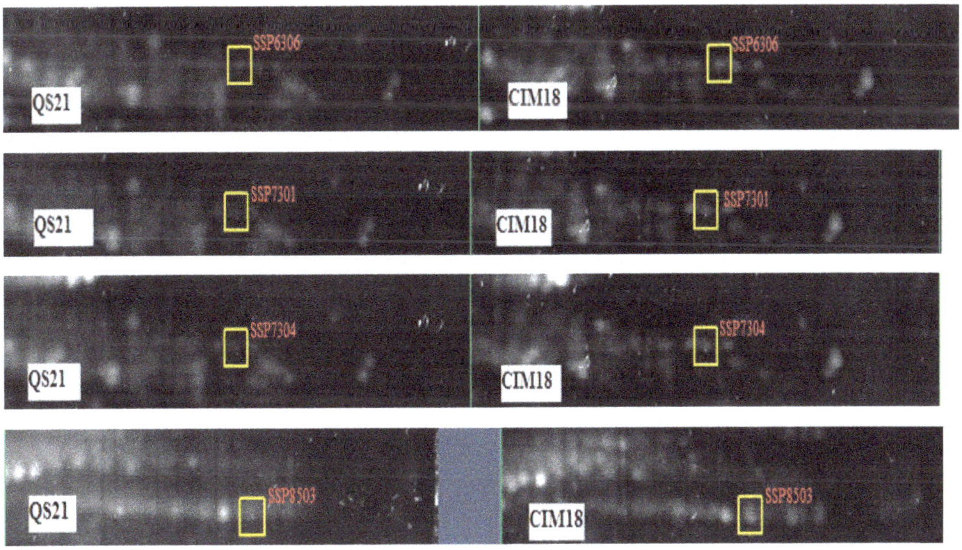

Figure A1. Gel images of protein spots occurring in maize leaves of both inbred lines, QS21 (susceptible) and CIM18 (tolerant), under combined drought and heat stress. Prominent spots indicate upregulation, and vice versa.

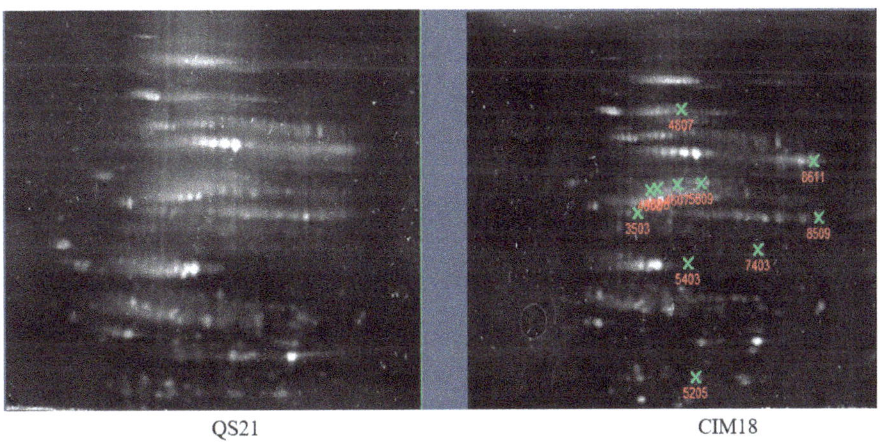

Figure A2. Representative 2-DE gel of susceptible QS21 and tolerant CIM18 lines under combined drought and heat stress. The green crosses show differentially expressed spots occurring in CIM18 but not in QS21.

Table A1. A two-way ANOVA of physiological and yield parameters measured under combined drought and heat stress treatments on inbred lines.

Parameters	Inbred	Treatment	Inbred Line × Treatment
Sub-stomatal CO_2	<0.001	NS	<0.001
Leaf temperature	<0.001	<0.001	<0.001
Photosynthetic yield	<0.001	<0.001	<0.001
Shoot	<0.001	<0.001	<0.001
Root	<0.001	<0.001	<0.001

NS—Not significant.

Table A2. Pedigree, drought tolerance and heterotic grouping of quality protein maize inbred lines exposed to combined drought and heat stress.

Line	Pedigree	Drought Tolerance	Heterotic Group
CIM1	(CLQRCWQ50/CML312SR)-2-2-1-BB-1-B-B	-	A
CIM2	[[CML202/CML144]F2-1-1-3-B-1-B*6/[GQL5/[GQL5/[MSRXPOOL9]C1F2-205-1(OSU23i)-5-3-X-X-1-BB]F2-4sx]-11-3-1-1-B*4]-B*5-1-B	drought-tolerant	B
CIM3	[CML141/[CML141/CML395]F2-1sx]-4-2-1-B*4-1-BB-B	drought-tolerant	B
CIM4	[CML144/[CML144/CML395]F2-5sx]-1-3-1-3-B*7-B	drought-tolerant	-
CIM5	[CML144/SNSYNF2[N3/TUX-A-90]-102-1-2-2-BSR-B*4]-B-4-3-B*4-1-B-B	drought-tolerant	B
CIM6	[CML150/CML373]-B-2-2-B*4-4-B-B	drought-tolerant	A
CIM7	[CML159/[CML159/[MSRXPOOL9]C1F2-205-1(OSU23i)-5-3-X-X-1-BB]F2-3sx]-8-1-1-BBB-4-B-B	drought-tolerant	A
CIM8	[CML182/TZMI703]-B-9-1-BB-#-BB-2-B-B	-	B
CIM9	[CML202/CML144]F2-1-1-3-B-1-B*6-2-B	drought-tolerant	B
CIM10	[CML205/CML176]-B-2-1-1-2-B*5-1-B-B	drought-tolerant	B
CIM11	[CML389/CML176]-B-29-2-2-B*4-B	drought-tolerant	B
CIM12	[GQL5/[GQL5/[MSRXPOOL9]C1F2-205-1(OSU23i)-5-3-X-X-1-BB]F2-4sx]-11-3-1-1-B*5-3-B-B	-	B
CIM13	[GQL5/[GQL5/CML202]F2-3sx]-11-4-1-3-B*4-B	-	B
CIM14	[TZMI703/CML176]-B-3-2-B*5-4-B-B	-	B
CIM15	CLQRCWQ50-BB-1-2-B-B	-	B
CIM16	CML176-#-B-2-B	drought-sensitive	B
CIM17	CML181-B-1-5-B-B7	drought-sensitive	-
CIM18	CML182-BB-B	-	-
CIM19	CML264Q-B-1-2-B-B	drought-sensitive	A
CIM20	CML491-B-3-11-B-B	-	A
CIM21	CML492-BB-2-1-B-B	drought-sensitive	B
QS1	K054W	-	F
QS4	V0548W	-	F
QS5	V0298W	-	F
QS6	B0388W	-	F
QS7	EM362W	-	M
QS8	EM583W	-	F
QS10	E625W	-	F
QS14	HM18W	-	O
QS15	HM233W	-	T
QS16	HM238W	-	M
QS17	HM267W	-	F
QS18	HM267W	-	F
QS19	HM268W	-	F
QS20	HM284W	-	H
QS21	HM1472W	-	B
QS22	JM226W	-	H
QS23	JM2341W	-	H
QS25	JM2561W	-	H

Table A2. *Cont.*

Line	Pedigree	Drought Tolerance	Heterotic Group
QS26	JM2602W		H
QS27	JM2621W	-	H
QS28	JM2641W	-	H
QS29	E5	-	G
QS30	E6	-	G
QS32	E27	-	G

CIM—CIMMTY; QS—Quality Seed; —unknown heterotic group; F—F2834T; M—M37W; O—Obatanpa; T—unknown Tropical; H—Hickory King; B—Brazil; G—Ghana; P—Potch Pearl; A—heterosis similar to Tuxepeno, N3, Reid and Kitale; B—heterosis similar to Ecuador, ETO, SC, Blanco and Lancaster.

References

1. Cui, D.; Wu, D.; Liu, J.; Li, D.; Xu, C.; Li, S.; Li, P.; Zhang, H.; Liu, X.; Jiang, C.; et al. Proteomic analysis of seedling roots of two maize inbred lines that differ significantly in the salt stress response. *PLoS ONE* **2015**, *10*, e0116697. [CrossRef]
2. Chávez-Arias, C.C.; Ramírez-Godoy, A.; Restrepo-Díaz, H. Influence of drought, high temperatures, and/or defense against arthropod herbivory on the production of secondary metabolites in maize plants. A review. *Curr. Plant Biol.* **2022**, *32*, 00268. [CrossRef]
3. Grzesiak, M.T.; Marcinska, I.; Janowiak, F.; Rzepka, A.; Hura, T. The relationship between seedling growth and grain yield under drought conditions in maize and triticale genotypes. *Acta Physiol. Plant* **2012**, *34*, 1757–1764. [CrossRef]
4. Millan, M.; Simonneau, T.; Coupe-Ledru, A.; Boulord, R.; Christophe, A.; Pallas, B. Relationships between leaf temperature, stomatal conductance and architecture: Potential impact on leaf burning among arange of genotypes in grapevine. *OENO ONE* **2023**, *57*, 345–359. [CrossRef]
5. Mahmood, U.; Li, X.; Fan, Y.; Chang, W.; Niu, Y.; Li, J.; Qu, C.; Lu, K. Multi-omics revolution to promote plant breeding efficiency. *Front. Plant Sci.* **2022**, *13*, 1062952. [CrossRef] [PubMed]
6. Wang, A.; Ma, C.; Ma, H.; Qiu, Z.; Wen, X. Physiological and proteomic responses of pitaya to PEG-induced drought stress. *Agriculture* **2021**, *11*, 632. [CrossRef]
7. Rollins, J.A.; Habte, E.; Templer, S.E.; Colby, T.; Schmidt, J.; von Korff, M. Leaf proteome alterations in the context of physiological and morphological responses to drought and heat stress in barley (*Hordeumvulgare* L.). *J. Exp. Bot.* **2013**, *64*, 3201–3212. [CrossRef]
8. Dong, A.; Yang, Y.; Liu, S.; Zenda, T.; Liu, X.; Wang, Y.; Li, J.; Duan, H. Comparative proteomics analysis of two maize hybrids revealed drought-stress tolerance mechanisms. *Biotechnol. Biotechnol. Equip.* **2020**, *34*, 63–780. [CrossRef]
9. Nemati, M.; Piro, A.; Norouzi, M.; Vahed, M.M.; Nisticò, D.M.; Mazzuca, S. Comparative physiological and leaf proteomic analyses revealed the tolerant and sensitive traits to drought stress in two wheat parental lines and their F6 progenies. *EEB* **2019**, *158*, 223–237. [CrossRef]
10. Lima, L.L.; Balbi, B.P.; Mesquita, R.O.; da Silva, J.C.F.; Coutinho, F.S.; Carmo, F.M.S.; Vital, C.E.; Mehta, A.; Loureiro, M.E.; Fontes, E.P.B.; et al. Proteomic and metabolomic analysis of a drought tolerant soybean cultivar from Brazilian savanna. *Crop Breed. Genet. Genom.* **2019**, *1*, e190022.
11. Boguszewska-Mankowska, D.; Gietler, M.; Nykiel, M. Comparative proteomic analysis of drought and high temperature response in roots of two potato cultivars. *J. Plant Growth Regul.* **2020**, *92*, 345–363. [CrossRef]
12. Dong, S.; Zhou, Q.; Yan, C.; Song, S.; Wang, X.; Wu, Z.; Wang, X.; Ma, C. Comparative protein profiling of two soybean genotypes with different stress tolerance reveals major components in drought tolerance. *Front. Sustain. Food Syst.* **2023**, *7*, 1200608. [CrossRef]
13. Younis, A.; Ramzan, F.; Ramzan, Y.; Zulfiqar, F.; Ahsan, M.; Lim, K.B. Molecular markers improve abiotic stress tolerance in crops: A review. *Plants* **2020**, *9*, 1374. [CrossRef]
14. Dos Santos, T.B.; Ribas, A.F.; de Souza, S.G.H.; Budzinski, I.G.F.; Domingues, D.S. Physiological responses to drought, salinity, and heat stress in plants: A review. *Stresses* **2022**, *2*, 113–135. [CrossRef]
15. Mondal, S.; Karmakar, S.; Panda, D.; Pramanik, K.; Bose, B.; Singhal, R.K. Crucial plant processes under heat stress and tolerance through heat shock proteins. *Plant Stress* **2023**, *10*, 100227. [CrossRef]
16. Asthir, B. Mechanisms of heat tolerance in crop plants. *Biol. Plant* **2015**, *59*, 620–628. [CrossRef]
17. San-Eufrasio, B.; Bigatton, E.D.; Guerrero-Sánchez, V.M.; Chaturvedi, P.; Jorrín-Novo, J.V.; Rey, M.D.; Castillejo, M.A. Proteomics data analysis for the identification of proteins and derived proteotypic peptides of potential use as putative drought tolerance markers for *Quercus ilex*. *Int. J. Mol. Sci.* **2021**, *22*, 3191. [CrossRef]
18. Xu, X.; Fonseca de Lima, C.F.; Vu, L.D.; De Smet, I. When drought meets heat–a plant omics perspective. *Front. Plant Sci.* **2023**, *14*, 1250878. [CrossRef] [PubMed]

19. Halder, T.; Choudhary, M.; Liu, H.; Chen, Y.; Yan, G.; Siddique, K.H.M. Wheat Proteomics for Abiotic Stress Tolerance and Root System Architecture: Current Status and Future Prospects. *Proteomes* **2022**, *10*, 17. [CrossRef]
20. Aina, O.; Bakare, O.O.; Fadaka, A.O.; Keyster, M.; Klein, A. Plant biomarkers as early detection tools in stress management in food crops: A review. *Planta* **2024**, *259*, 60. [CrossRef]
21. Sehgal, A.; Sita, K.; Kumar, J.; Kumar, S.; Singh, S.; Siddique, K.H.; Nayyar, H. Effects of drought, heat and their interaction on the growth, yield and photosynthetic function of lentil (*Lens culinaris* Medikus) genotypes varying in heat and drought sensitivity. *Front. Plant Sci.* **2017**, *8*, 1776. [CrossRef]
22. Marchin, R.M.; Backes, D.; Ossola, A.; Leishman, M.R.; Tjoelker, M.G.; Ellsworth, D.S. Extreme heat increases stomatal conductance and drought-induced mortality risk in vulnerable plant species. *Glob. Change Biol.* **2022**, *3*, 1133–1146. [CrossRef]
23. Avramova, V.; Nagel, K.A.; AbdElgawad, H.; Bustos, D.; DuPlessis, M.; Fiorani, F.; Beemster, G.T. Screening for drought tolerance of maize hybrids by multi-scale analysis of root and shoot traits at the seedling stage. *J. Exp. Bot.* **2016**, *67*, 2453–2466. [CrossRef] [PubMed]
24. Pfunde, C.N. Genetics, Physiology, Proteomics of Quality Protein Maize Inbred Lines Under Drought and Heat Stress. Ph.D. Thesis, University of Fort Hare, Alice, South Africa, May 2017.
25. Subramani, M.; Urrea, C.A.; Kalavacharla, V. Comparative Analysis of Untargeted Metabolomics in Tolerant and Sensitive Genotypes of Common Bean (*Phaseolus vulgaris* L.) Seeds Exposed to Terminal Drought Stress. *Metabolites* **2022**, *12*, 944. [CrossRef]
26. Meng, X.; Liu, S.; Dong, T.; Xu, T.; Ma, D.; Pan, S.; Li, Z.; Zhu, M. Comparative Transcriptome and Proteome Analysis of Salt-Tolerant and Salt-Sensitive Sweet Potato and Overexpression of IbNAC7 Confers Salt Tolerance in Arabidopsis. *Front. Plant Sci.* **2020**, *11*, 572540. [CrossRef] [PubMed]
27. Laurindo, B.S.; Laurindo, R.D.F.; Fontes, P.P.; Vital, C.E.; Delazari, F.T.; Baracat-Pereira, M.C.; da Silva, D.J.H. Comparative analysis of constitutive proteome between resistant and susceptible tomato genotypes regarding to late blight. *Funct. Integr. Genom.* **2018**, *18*, 11–21. [CrossRef] [PubMed]
28. Ghirardo, A.; Nosenko, T.; Kreuzwieser, J.; Winkler, J.B.; Kruse, J.; Albert, A.; Merl-Pham, J.; Lux, T.; Ache, P.; Zimmer, I.; et al. Protein expression plasticity contributes to heat and drought tolerance of date palm. *Oecologia* **2021**, *197*, 903–919. [CrossRef] [PubMed]
29. Das, A.; Eldakak, M.; Paudel, B.; Kim, D.W.; Hemmati, H.; Basu, C.; Rohila, J.S. Leaf proteome analysis reveals prospective drought and heat stress response mechanisms in soybean. *Biomed. Res. Int.* **2016**, *1*, 6021047. [CrossRef]
30. Abdulbaki, A.S.; Alsamadany, H.; Alzahrani, Y.; Olayinka, B.U. Rubisco and abiotic stresses in plants: Current assessment. *Turk. J. Bot.* **2022**, *46*, 541–552. [CrossRef]
31. Loka, D.A.; Oosterhuis, D.M.; Baxevanos, D.; Noulas, C.; Hu, W. Single and combined effects of heat and water stress and recovery on cotton (*Gossypium hirsutum* L.) leaf physiology and sucrose metabolism. *Plant Physiol. Biochem.* **2020**, *148*, 166–179. [CrossRef]
32. Vanlerberghe, G.C. Alternative oxidase: A mitochondrial respiratory pathway to maintain metabolic and signaling homeostasis during abiotic and biotic stress in plants. *Int. J. Mol. Sci.* **2013**, *14*, 6805–6847. [CrossRef] [PubMed]
33. Wang, J.; Gao, H.; Guo, Z.; Meng, Y.; Yang, M.; Li, X.; Yang, Q. Adaptation responses in C4 photosynthesis of sweet maize (*Zea mays* L.) exposed to nicosulfuron. *Ecotoxicol. Environ. Saf.* **2021**, *214*, 112096. [CrossRef] [PubMed]
34. Li, X.; Dan, X.; Liu, J.; Liv, Q.; Li, X. Translation elongation factor-1α is pivotal for plant heat tolerance despite its pronounced heat-induced aggregation. *Plant Physiol. Biochem.* **2024**, *210*, 108649. [CrossRef]
35. Rahman, M.A.; Woo, J.H.; Song, Y.; Lee, S.H.; Hasan, M.M.; Azad, M.A.K.; Lee, K.W. Heat shock proteins and antioxidant genes involved in heat combined with drought stress responses in perennial rye grass. *Life* **2022**, *12*, 1426. [CrossRef]
36. Klein, R.D.; Chidawanyika, T.; Tims, H.S.; Meulia, T.; Bouchard, R.A.; Pett, V.B. Chaperone function of two small heat shock proteins from maize. *Plant Sci.* **2014**, *221*, 48–58. [CrossRef]
37. Dharan, S.S.; Sabu, K.K. Expression profiling of stress responsive genes in cell suspension of *Elettaria cardamomum* (L.) Maton under abiotic stress. *J. Plant. Crops.* **2023**, *51*, 23–30. [CrossRef]
38. Yusof, N.A.; Masnoddin, M.; Charles, J.; Thien, Y.Q.; Nasib, F.N.; Michael, C.; Wong, V.L.; Murad, A.M.A.; Mahadi, N.M.; Bharudin, I. Can heat shock protein 70 (HSP70) serve as biomarkers in Antarctica for future ocean acidification, warming and salinity stress? *Polar Biol.* **2022**, *45*, 371–394. [CrossRef]
39. Zhang, L.; Zhao, H.K.; Dong, Q.L.; Zhang, Y.Y.; Wang, Y.M.; Li, H.Y.; Xing, G.J.; Li, Q.Y.; Dong, Y.S. Genome-wide analysis and expression profiling under heat and drought treatments of HSP70 gene family in soybean (*Glycine max* L). *Front. Plant Sci.* **2015**, *6*, 773. [CrossRef]
40. Raja, V.; Qadir, S.U.; Alyemeni, M.N.; Ahmad, P. Impact of drought and heat stress individually and in combination on physio-biochemical parameters, antioxidant responses, and gene expression in *Solanum lycopersicum*. *3 Biotech* **2020**, *10*, 208. [CrossRef]

41. Yousaf, M.I.; Riaz, M.W.; Jiang, Y.; Yasir, M.; Aslam, M.Z.; Hussain, S.; Sajid Shah, S.A.; Shehzad, A.; Riasat, G.; Manzoor, M.A.; et al. Concurrent effects of drought and heat stresses on physio-chemical attributes, antioxidant status and kernel quality traits in maize (*Zea mays* L.) hybrids. *Front. Plant Sci.* **2022**, *13*, 898823. [CrossRef] [PubMed]
42. Zhang, Z.W.; Feng, L.Y.; Cheng, J.; Tang, H.; Xu, F.; Zhu, F.; Zhao, Z.Y.; Yuan, M.; Chen, Y.E.; Wang, J.H.; et al. The roles of two transcription factors, ABI4 and CBFA, in ABA and plastid signalling and stress responses. *Plant Mol. Biol.* **2013**, *83*, 445–458. [CrossRef] [PubMed]
43. Hasan, M.S.; Singh, V.; Islam, S.; Islam, M.S.; Ahsan, R.; Kaundal, A.; Islam, T.; Ghosh, A. Genome-wide identification and expression profiling of glutathione S-transferase family under multiple abiotic and biotic stresses in *Medicago truncatula* L. *PLoS ONE* **2021**, *16*, 0247170. [CrossRef] [PubMed]
44. Kumar, S.; Trivedi, P.K. Glutathione S-transferases role in combating abiotic stresses including arsenic detoxification in plants. *Front. Plant Sci.* **2018**, *9*, 751. [CrossRef] [PubMed]
45. Zheng, H.; Yang, Z.; Wanga, W.; Guod, S.; Lic, Z.; Liuc, K.; Suia, N. Transcriptome analysis of maize inbred lines differing in drought tolerance provides novel insights into the molecular mechanisms of drought responses in roots. *Plant Physiol. Biochem.* **2020**, *149*, 11–26. [CrossRef] [PubMed]
46. Sun, Y.; Shi, M.; Wang, D.; Gong, Y.; Sha, Q.; Liv, P.; Yang, J.; Chu, P.; Guo, S. Research progress on the roles of actin-depolymerizing factor in plant stress responses. *Front. Plant Sci.* **2023**, *14*, 1278311. [CrossRef]
47. Sniegowska-Swierk, K.; Dubas, E.; Rapacz, M. Drought-induced changes in the actin cytoskeleton of barley (*Hordeum vulgare* L.) leaves. *Acta Physiol. Plant* **2015**, *37*, 73. [CrossRef]
48. Bawa, G.; Liu, Z.; Zhou, Y.; Fan, S.; Ma, Q.; Tissue, D.T.; Sun, X. Cotton proteomics: Dissecting the stress response mechanisms in cotton. *Front. Plant Sci.* **2022**, *13*, 1035801. [CrossRef]
49. Hu, X.; Wu, L.; Zhao, F.; Zhang, D.; Li, N.; Zhu, G.; Li, C.; Wang, W. Phosphoproteomic analysis of the response of maize leaves to drought, heat and their combination stress. *Front. Plant Sci.* **2015**, *6*, 298. [CrossRef]
50. Nieto-Sotelo, J.; Martinez, L.; Ponce, G.; Cassab, G.I.; Alagon, A.; Meely, R.B.; Ribaut, J.; Yang, R. Maize HSP101 plays important roles in both induced and basal thermotolerance and primary radicle growth. *Plant Cell* **2002**, *14*, 1621–1633. [CrossRef]
51. Farshadfar, E.; Poursiahbidi, M.M.; Safavi, S.M. Assessment of drought tolerance in land races of bread wheat based on resistance/tolerance indices. *Int. J. Adv. Biol. Biomed. Res* **2013**, *1*, 14–158.

plants

MDPI

Article

Genetic Gains and Field Validation of Synthetic Populations in Tropical Maize Using Selection Indexes and REML/BLUP

Antônia Maria de Cássia Batista de Sousa [1], Marcela Pedroso Mendes Resende [1], Ailton Jose Crispim-Filho [1], Glauco Vieira Miranda [2] and Edésio Fialho dos Reis [3,*]

[1] Escola de Agronomia, Universidade Federal de Goiás, Goiânia 74690-900, Brazil; antonia.sousaufpi@gmail.com (A.M.d.C.B.d.S.); marcelapmr@ufg.br (M.P.M.R.); ailton.crispimfilho@gmail.com (A.J.C.-F.)

[2] Agronomy Coordination, Federal Technological University of Paraná, Santa Helena 85892-000, Brazil; glaucovmiranda@utfpr.edu.br

[3] Instituto de Biociências, Universidade Federal de Jataí, Jataí 75801-615, Brazil

* Correspondence: edesioreis@ufj.edu.br

Abstract: The development of tropical maize populations with high heterosis potential is essential for sustaining genetic progress in hybrid breeding programs, yet accurate selection remains challenging due to genotype–phenotype interactions and inbreeding depression. This study evaluated the efficiency of five selection strategies in recurrent selection programs using F_2 populations derived from commercial maize hybrids: Smith–Hazel Index (SHI), Base Index (BIA), Mulamba–Mock Index (MMI), REML/BLUP for grain yield (BLUP_GY), and REML/BLUP for inbreeding depression (BLUP_ID). Consistency among methods was assessed with a heatmap, and predicted genetic gains were compared with realized field performance. Predicted gains were highest with MMI and BIA for grain yield and ear weight, although realized results revealed discrepancies, particularly for BLUP-based approaches. Notably, BLUP_GY, which had the lowest predicted yield (4025 kg ha^{-1}), achieved a realized yield of 5620 kg ha^{-1}, surpassing BIA and SHI. This indicates that additive potential was underestimated in predictions, likely due to dominance and environmental effects in early F_2 cycles. Overall, BLUP-based methods proved effective in identifying progenies with higher additive value, and their integration with phenotypic indices is recommended to combine short-term realized gains with sustained genetic improvement.

Keywords: *Zea mays*; inbreeding depression; heterosis

1. Introduction

The development of new tropical maize (*Zea mays* L.) populations with high heterosis potential is a central objective in hybrid breeding programs, especially in the face of climate change and the rising global demand for food. Continuous genetic progress is essential for these programs and relies on efficient strategies to accurately identify superior progenies and form high-performing heterotic groups. However, selection in early recurrent cycles remains challenging due to the complex interaction between genotypic and phenotypic components, further complicated by inbreeding depression.

Traditional phenotypic selection methods, such as mass selection or classical selection indexes, have historically played an important role in breeding pipelines. Nevertheless, their effectiveness is often constrained by environmental variability and their limited ability to account for genetic relationships and additive variance [1]. In this context, modern

 https://doi.org/10.3390/plants14203149

statistical approaches—especially mixed models based on Restricted Maximum Likelihood (REML) and Best Linear Unbiased Prediction (BLUP)—have become increasingly prominent due to their capacity to generate accurate genotypic estimates under unbalanced designs and across multi-environment trials [2,3].

Several studies have highlighted the advantages of using REML/BLUP in tropical maize breeding. The mixed models could explain more than 99% of phenotypic variance in key traits while maintaining stability under genotype-by-environment interactions [3]. The predictive performance is retained even with up to 20% of hybrids or 23% of environments missing, attesting to the robustness of the methodology [2]. Pedigree-based BLUP led to superior genetic gains in popcorn expansion volume (1–45%), outperforming mass selection [4].

The potential of REML/BLUP in driving genetic gains in tropical maize has been confirmed in recent studies, with improvements of up to 24.07% in traits such as ear length and grain weight in half-sib progenies [5]. Similarly, annual gains between 46 and 118 kg ha^{-1} year^{-1} in early-maturing maize hybrids developed by CMMIYT across 68 environments in seven African countries have been documented, highlighting the effectiveness of structured breeding programs and robust predictive tools in diverse and stress-prone conditions [6].

REML/BLUP outperformed classical selection indexes in predicting genetic gains and estimating genotypic values in supersweet corn programs [7]. While both REML/BLUP and the Mulamba and Mock index showed high coincidence under balanced conditions, REML/BLUP demonstrated superior gains and reliability in multi-trait selection. However, in low-resource settings or early recurrent selection cycles with limited data, simpler phenotypic selection methods remain relevant, with the Mulamba and Mock index standing out for its simplicity, flexibility, and consistent performance across crops such as maize and sugarcane [1,8].

Advances in quantitative genetics, statistical modeling, and molecular tools have significantly improved maize breeding efficiency. The integration of genomic selection (GS) with strategies tailored to trait heritability and genetic architecture has been shown to accelerate genetic gains while preserving genetic diversity [9]. Reciprocal recurrent genomic selection (RRGS) has emerged as a powerful approach to improve both general and specific combining abilities. Comparative analyses between full-sib and half-sib training sets reveal that full-sib strategies often result in higher cumulative selection gains, particularly when specific combining ability (SCA) plays a prominent role [10,11].

Genetic diversity remains a cornerstone of successful breeding programs. Studies using SSR markers in convergent-derived maize populations and in introgressed progenies from wild relatives like *Zea nicaraguensis* emphasize the importance of broadening the genetic base to enhance selection response and population resilience [12,13]. In Brazil, substantial genetic gains have been achieved through intrapopulation recurrent selection, with high heritability and consistent improvement across five selection cycles in fresh corn half-sib progenies, and confirmation of the genetic potential of southern Brazilian maize populations even under low-input conditions [14,15].

Reciprocal recurrent selection with full-sib progenies has also been effectively applied to enhance both heterosis and parental line development, consistent with theoretical models highlighting the benefits of optimizing training sets, maintaining genetic variability, and balancing short- and long-term selection responses [10,16,17]. Thus, integrating REML/BLUP with classical selection indexes—and potentially with ensemble learning or deep learning models—represents a promising strategy for modern breeding optimization [18].

This study builds upon these advancements by comparing the efficiency of three classical selection indexes (Smith–Hazel, Base Index, and Mulamba–Mock) and the REML/BLUP methodology for grain yield and inbreeding depression in selecting 194 half-sib progenies and their S1 progenies, focusing on grain yield and inbreeding depression. By evaluating predicted and realized genetic gains after recombination, the objective is to identify the most effective strategies for constructing genetically diverse base populations with strong heterotic potential for tropical maize. Ultimately, this research addresses the need for accurate, scalable, and efficient selection methods to support the rational formation of high-performing synthetic populations adapted to tropical environments. In doing so, it contributes to sustainable hybrid maize breeding through the integration of quantitative genetics, advanced statistical modeling, and field-based experimentation.

2. Results

2.1. Selected Progenies for Grain Yield and Inbreeding Depression

The heatmap illustrates the overlap of selected maize progenies among five different selection methods, BIA, MMI, SHI, BLUP_GY, and BLUP_ID, revealing meaningful patterns of convergence and divergence that provide insights into the consistency and complementarity of phenotypic and genotypic approaches in early recurrent selection cycles (Figure 1). A consistent pattern was observed for progenies 22, 39, 43, 135, and 152, which were selected simultaneously by all methods, indicating strong agreement among the different approaches. Progenies such as 101, 113, and 137 were selected by four methods, reflecting high but not complete convergence. In contrast, progenies like 145, 19, 17, and 177 were identified by only two methods, demonstrating limited consensus. A considerable number of progenies were exclusively selected by BLUP_GY and BLUP_ID, without overlap with the index-based strategies (SHI, BIA, and MMI). Additionally, some progenies, including 35, 89, 90, 131, 134, 147, and 168, appeared in only one method, revealing unique selections not shared across approaches. Overall, the results reveal the presence of both widely recurrent progenies across all selection strategies and method-specific progenies, evidencing distinct patterns of consistency and divergence among the applied methodologies.

The coincidence index heatmap shows the degree of overlap in the top 10 selected maize progenies among five different selection methods (Figure 2). The results show a strong agreement between the genomic selection strategies BLUP_GY and BLUP_ID (coincidence index = 0.82), suggesting consistency in the ranking of top-performing progenies based on genomic predictions. In contrast, classical phenotypic index-based methods such as SHI, BIA, and MMI exhibited moderate to low overlap among themselves (e.g., BIA vs. MMI = 0.67; SHI vs. BIA = 0.33) and no overlap with BLUP-based approaches (coincidence index = 0.00 for all comparisons between BLUP_GY or BLUP_ID with SHI, BIA, or MMI).

The heatmap using the top 20 progenies prioritized by each method presents the coincidence index calculated between five selection strategies (Figure 3). The genomic methods BLUP_GY and BLUP_ID exhibited a high concordance (coincidence index = 0.82), confirming strong consistency between their selection outputs. In contrast, phenotypic index-based strategies (SHI, BIA, MMI) showed moderate agreement among themselves—especially BIA vs. MMI (0.74) and SHI vs. MMI (0.48)—while exhibiting minimal overlap with the genomic strategies (only 0.05 in all comparisons between BLUP_GY or BLUP_ID and SHI, BIA, or MMI).

To evaluate the convergence trend among strategies, we calculated the average pairwise coincidence index as a function of the number of top-ranked progenies. The average index increased steadily from 0.23 (top 10) to 0.39 (top 40).

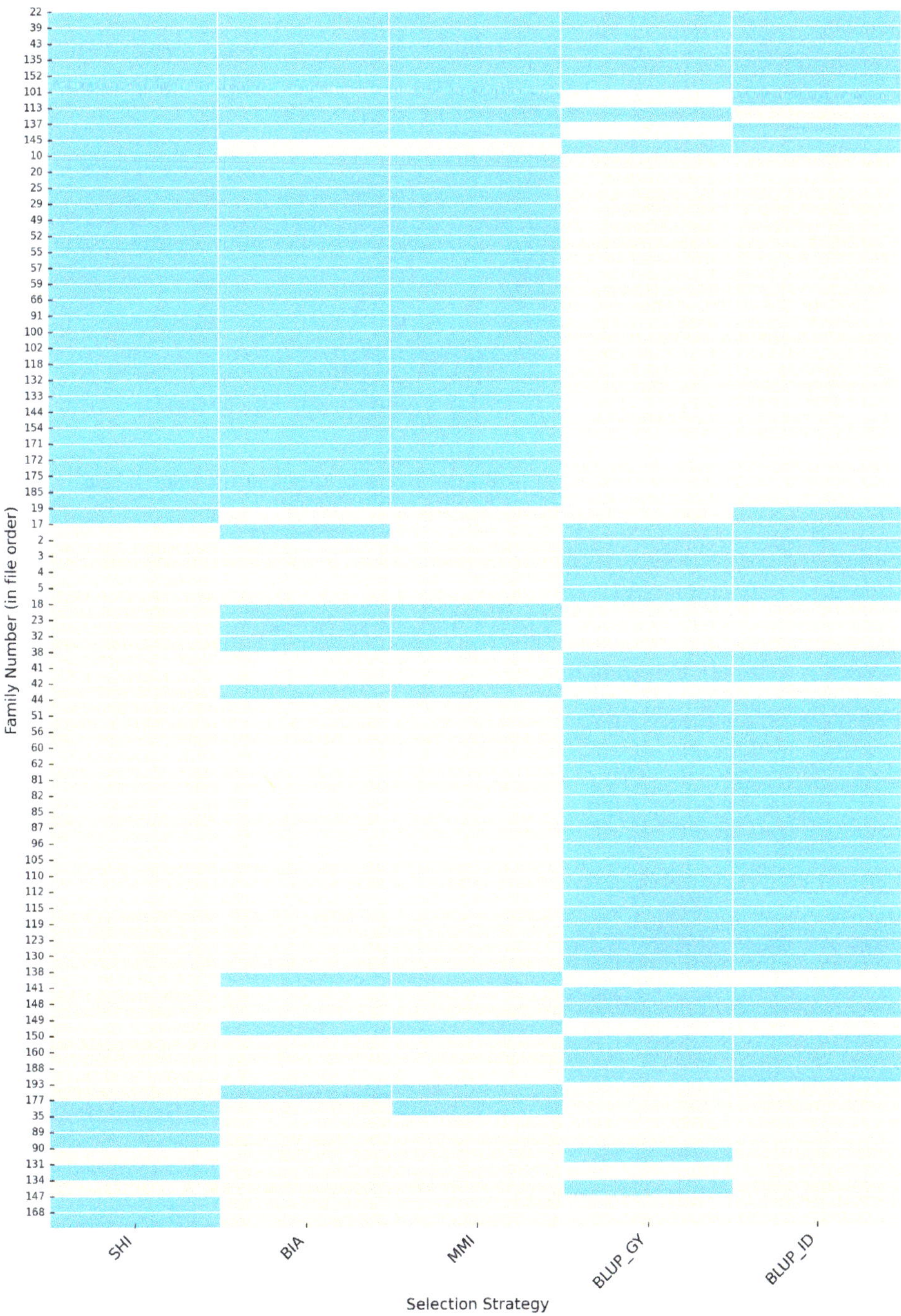

Figure 1. Heatmap of the maize progenies selected by five different breeding strategies: BIA (Base Index Approach), SHI (Smith–Hazel Index), MMI (Mulamba–Mock Index), BLUP_GY (REML/BLUP for Grain Yield), and BLUP_ID (REML/BLUP for Inbreeding Depression). Each cell indicates whether a specific line was selected by the corresponding method (blue = selected; yellow = not selected).

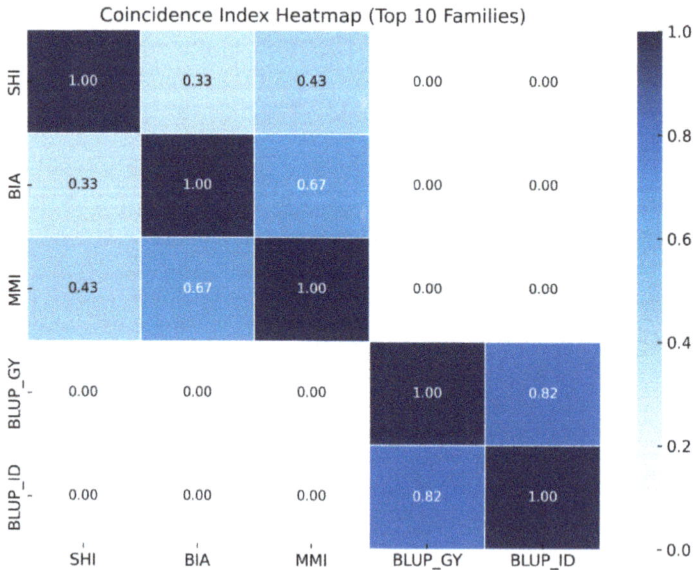

Figure 2. Heatmap shows the coincidence index among five selection methods—BIA (Base Index Approach), SHI (Smith–Hazel Index), MMI (Mulamba–Mock Index), BLUP_GY (REML/BLUP for Grain Yield), and BLUP_ID (REML/BLUP for Inbreeding Depression)—based on the top 10 most frequently selected maize progenies. Values represent the proportion of progenies jointly selected by each pair of methods.

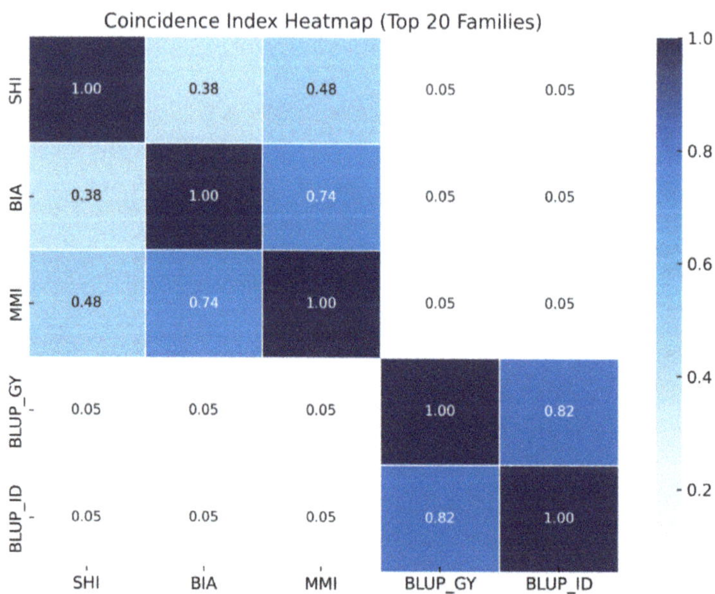

Figure 3. Heatmap presents the coincidence index among five different selection strategies for tropical maize based on the 20 most frequently selected progenies. It highlights the consistency or divergence in line selection among phenotypic indexes (BIA, SHI, MMI) and REML/BLUP-based methods (BLUP_GY and BLUP_ID). Higher values indicate greater agreement in line selection, supporting integrated use or comparative assessment of these approaches in breeding pipelines.

2.2. Predicted Selection Gains Based on Phenotype Indexes

The greatest predicted genetic gains with 20% of selected progenies were observed for grain yield (GY) and ear weight (EW), especially using the MMI and BI indexes (Table 1). The MMI index predicted the gain for grain yield (15.93%), followed closely by BI (15.84%) and SHI (14.79%). For ear weight, gains were also substantial, with values ranging from 13.41% (SHI) to 14.01% (BI). Inbreeding depression (ID), considered a key trait in this analysis, exhibited strong negative values across all indexes (SHI: −14.45%, BI: −14.96%, MMI: −15.07%).

Table 1. Predicted selection gains (PG) of 38 selected progenies using the Smith–Hazel index (SHI), Base index (BI), and Mulamba–Mock Index (MMI) in 194 half-sib maize progenies.

Variable	SHI_Xs	SHI (%)	BI_Xs	BI (%)	MMI_Xs	MMI (%)
GY	4515	14.79	4597	15.84	4590	15.93
EW	5480	13.41	5534	14.01	5531	13.98
ID	36.44	−14.45	35.93	−14.96	35.82	−15.07
NE	14.92	7.56	14.62	6.19	14.61	6.11
ED	14.72	4.72	14.48	3.61	14.52	3.79
EL	4.40	2.17	4.37	1.75	4.36	1.71
MF	59.71	0.24	59.62	0.15	59.62	0.16
FF	58.50	−0.05	58.34	−0.19	58.42	−0.12
PH	2.250	−0.05	2.26	−0.05	2.25	−0.06
EH	1.010	−0.09	1.01	−0.07	1.01	−0.07

GY: grain yield (kg ha^{-1}); EW: ear weight (kg ha^{-1}); ID: inbreeding depression (% relative to the base population); NE: number of ears; ED: ear diameter (cm); EL: ear length (cm); MF: male flowering (days); FF: female flowering (days); PH: plant height (m); EH: ear height (m).

Moderate selection gains were recorded for traits such as number of ears per plant (NE), ear length (EL), and ear diameter (ED), with NE showing the most consistent response across methods—SHI: 7.56%, BI: 6.19%, and MMI: 6.11%, confirming its capacity to balance multiple traits when no single trait dominates the selection objective.

In contrast, flowering traits (female and male flowering, FF and MF), plant height (PH), and ear height (EH) exhibited negligible gains across all indexes, indicating limited selection pressure or unfavorable correlations with the main selection target (grain yield) suggesting a favorable trend toward more compact plants, which is desirable in tropical environments for lodging resistance and mechanical harvesting. For example, female flowering (FF) showed negative gains in all cases, ranging from −0.05% (SHI) to −0.19% (BI), suggesting a potential delay in flowering as a correlated response, which may contribute to reduced crop cycle and better adaptation to the second season.

2.3. Prediction of Genotypic Values

The evaluation of grain yield in 194 half-sib maize progenies from the third progeny generation (3G) using REML/BLUP methodology revealed substantial variability in both the predicted additive genotypic values ($\mu_i + g_i$) and associated inbreeding depression ($u_i + g_i$) (Table 2).

Among the 38 selected progenies, progeny 135 ranked at the top by point estimates for predicted grain yield ($\mu_i + g_i = 4890$ kg ha^{-1}) and BLUP (1423 kg ha^{-1}), showed low inbreeding depression (33.7), and was selected by all methods, with no statistically significant differences versus other top entries (Table 2; Figure 1).

Notably, the high predicted additive values ($\mu_i + g_i$) and low inbreeding depression observed in selected progenies, such as 135 and 145, demonstrate that modern breeding methods can recover and enhance yield potential without compromising genetic stability.

Table 2. Predicted additive genotypic values ($\mu_i + g_i$) for grain yield and inbreeding depression (%) across three generations of main half-sib maize progenies for PopBLUP_GY, and BLUP_ID.

3rd G	Blup_GY	$\mu_i + g_i$	Blup_ID	$u_i + g_i$
135	1423	4890	−17.3	33.7
145	1059	4525	−15.8	35.2
85	1020	4486	−10.9	40.1
150	746	4212	−11.2	39.8
87	706	4173	−10.6	40.4
4	626	4092	−9.4	41.7
44	604	4071	−9.1	41.9
82	580	4047	−8.6	42.4
96	568	4035	−8.5	42.5
105	565	4031	−8.4	42.6
Mean	558	4025	−7.9	43.1

Similarly, progeny 145 demonstrated a predicted yield of 4525 kg ha^{-1} and a modest inbreeding depression (35.2), reinforcing its potential for selection. Other high-performing progenies, such as 85 (4486 kg ha^{-1}) and 150 (4212 kg ha^{-1}), also ranked among the best in terms of grain yield; however, their inbreeding depression values were relatively higher (40.10 and 39.8, respectively).

Conversely, some progenies exhibited consistently lower performance, such as 26, 32, 43, and 7 recorded BLUPs below 410 kg ha^{-1} and genotypic values ($\mu_i + g_i$) under 3900 kg ha^{-1}, suggesting limited contribution to yield improvement. These genotypes may be excluded from further breeding unless they display desirable traits for other agronomic purposes.

The correlation between BLUPs and predicted additive values was positive, indicating that higher phenotypic expression is generally associated with greater additive genetic merit. This is reflected in the clustering of selected progenies above the average thresholds for both parameters (BLUP = 558 kg ha^{-1}; $\mu_i + g_i$ = 4025 kg ha^{-1}).

Furthermore, the evaluation across three breeding generations (1st, 2nd, and 3rd) reinforces the heritability and stability of grain yield traits. Genotypes such as 135, 145, and 85, which maintained superior performance across generations, are particularly valuable for integration into long-term breeding strategies.

The average predicted yield for the 38 selected progenies was 4025 kg ha^{-1}, with a corresponding average inbreeding depression of 43.1, and a predicted genetic gain of 558 kg ha^{-1} for grain yield and a reduction of 7.90 units for inbreeding depression. The simultaneous consideration of both parameters, high genotypic values for yield and low inbreeding depression, allowed the identification of nine recurrent progeny (135, 145, 85, 150, 87, 4, 44, 42, 96 and 105) that showed consistent performance across generations. Notably, these progenies originated from first-generation progenies previously selected for resistance to foliar diseases, such as gray leaf spot. This convergence highlights the possibility that such progenies carry favorable alleles for both yield and stress tolerance, enhancing their breeding value and justifying their recombination in population improvement programs.

In summary, the results confirm the genetic superiority of a core set of progenies within RV-02, especially those descending from progeny 23, reinforcing the potential of this subgroup as the foundation for developing high-performing, genetically stable tropical maize populations.

2.4. *Synthetic Populations Performance*

The comparison of means among five maize populations developed via different selection strategies, along with two commercial checks (hybrid P3898 and open-pollinated variety AL Bandeirante), revealed significant genetic progress for grain yield (GY) and other yield-related traits (Table 3).

Table 3. Comparison of means for male flowering (MF), ear length (EL), ear weight (EW), and grain yield (GY) evaluated in five maize populations derived from different selection strategies (PopBLUP_GY, PopBIA, PopSHI, PopMMI, and PopBLUP_ID), along with two commercial cultivars (Hybrid P3898; open-pollinated variety AL Bandeirante).

Populations	MF Days	EL cm	EW (kg ha^{-1})	GY (kg ha^{-1})	Predict GY (kg ha^{-1})
P3898	67.83 a *	18.38 ab	11,620 a	9400 a	
PopMMI	66.67 ab	18.88 a	7880 b	6300 b	4590
PopBLUP_GY	66.33 ab	16.33 ab	7060 b	5620 b	4025
PopBIA	65.33 b	15.88 ab	5980 c	4760 c	4597
PopSHI	66.33 ab	17.17 ab	5900 c	4650 c	4515
PopBLUP_ID	66.33 ab	15.75 b	6100 c	4810 c	
AL Bandeirantes	66.67 ab	18.04 ab	6010 c	4480 c	

* Means followed by the same letter in a column do not differ significantly by Tukey's test at 5% probability.

Among the synthetic populations, PopMMI and PopBLUP_GY demonstrated the highest agronomic performance, exhibiting statistically superior ear weight and grain yield, significantly outperforming the open-pollinated check (AL Bandeirante, GY = 4.48 t ha^{-1}). This yield surpassed that of the base populations (PopSHI, PopBIA, and PopBLUP_ID) and remained statistically inferior to the commercial check P3898 (9.40 t ha^{-1}). Although not statistically equal to P3898, the BLUP_GY population showed evidence of additive genetic gain, validating the use of mixed-model approaches in early-stage recurrent selection cycles. On the other hand, PopBIA, PopSHI, and PopBLUP_ID exhibited lower grain yield and ear weight values, ranging from 4.65 to 4.81 t ha^{-1} for GY and 5.90 to 6.10 t ha^{-1} for EW. Despite being inferior to PopMMI and PopBLUP_GY, these populations still performed comparably or better than the open-pollinated check, indicating partial gain through selection. However, further cycles of recurrent selection or incorporation of genomic tools may be required to boost their yield potential.

Male flowering (MF) did not differ significantly among most populations, except for P3898, which was slightly later (67.83 days) than the experimental groups (65.33 to 66.67 days). This suggests that selection did not strongly impact flowering time, allowing gains in yield without major shifts in phenology.

A comparison between predicted grain yield (Predict GY) and the realized values of grain yield (GY) revealed important discrepancies among the populations evaluated. Among the synthetic populations derived from selection indices, the predicted genetic gains were consistently higher than the realized values. For example, PopMMI presented a predicted GY of 4590 kg ha^{-1}, whereas the realized value was 6300 kg ha^{-1}, indicating that the prediction underestimated the actual performance. A similar pattern occurred with PopBIA (predicted 4597 vs. realized 4760 kg ha^{-1}) and PopSHI (predicted 4515 vs. realized 4650 kg ha^{-1}). In these cases, although the absolute difference was modest, the realized values slightly exceeded the predictions, suggesting that the additive genetic effects captured by the indices were expressed more favorably under the experimental conditions.

Conversely, for PopBLUP_GY and PopBLUP_ID, the realized yields were 5620 and 4810 kg ha^{-1}, respectively, while the predictions were 4025 and not available for BLUP_ID. Here, the differences were larger, particularly in PopBLUP_GY, where realized GY was substantially higher than predicted. This divergence may be attributed to environmental contributions or non-additive effects (dominance, epistasis) not fully accounted for by the BLUP-based predictions.

The comparison between predicted (Predict GY) and realized GY values highlights consistent differences across the evaluated populations. In general, the realized yields exceeded the predicted values, although the magnitude of the difference varied depending on the selection strategy. For instance, PopMMI showed a realized GY of 6300 kg ha^{-1}, compared with a predicted yield of 4590 kg ha^{-1}, while PopBIA and PopSHI also exhibited higher realized values (4760 and 4650 kg ha^{-1}) relative to their respective predictions (4597 and 4515 kg ha^{-1}).

In contrast, the populations based on BLUP methodologies presented distinct outcomes. PopBLUP_GY had a predicted yield of 4025 kg ha^{-1}, whereas the realized yield reached 5620 kg ha^{-1}, a substantial difference of more than 1500 kg ha^{-1}, indicating that additive effects captured in the prediction were sufficient to fully explain the realized performance under field conditions. PopBLUP_ID, on the other hand, did not include a predicted yield in the dataset, but the realized value was 4810 kg ha^{-1}, positioning it slightly below PopBLUP_GY in terms of actual performance.

3. Discussion

The overlap of selected progenies among the five strategies provides important insights into the consistency and divergence of phenotypic and BLUP-based approaches in early recurrent selection. The recurrent identification of progenies 22, 39, 43, 135, and 152 across all methods suggests that these progenies combine both favorable phenotypic performance and stable predicted genetic merit, reinforcing their potential as reliable candidates for advancement in breeding pipelines.

Overall, the combination of consistent and method-specific selections emphasizes the complementarity of phenotypic indices and BLUP-based approaches, reinforcing the importance of integrating multiple strategies to increase the robustness of selection decisions. This integration can maximize genetic gain by balancing convergence on superior and stable progeny with the exploration of unique candidates that expand the genetic base for long-term breeding progress.

In line with previous studies, the MMI showed strong performance in concentrating desirable traits and suppressing undesirable ones [1,19]. The consistent reductions in flowering time, plant height, and inbreeding depression, alongside improvements in yield components, indicate that this index is highly suitable for simultaneous multi-trait selection in maize half-sib populations under tropical conditions.

The average pairwise coincidence index as a function of the number of top-ranked progenies (top 10 to top 40) indicates that as more progenies are included in the ranking, the level of agreement among the selection strategies improves. However, even at the top 40 threshold, the average index remains below 0.40, emphasizing substantial methodological differences in selection criteria. Phenotypic indices tend to prioritize progenies with superior agronomic traits under field conditions, while genomic approaches select based on predicted genetic values and underlying variance components. This divergence reinforces the importance of integrating both phenotypic and genomic information to improve the robustness of selection decisions in early-stage recurrent selection cycles.

These findings highlight the utility of coincidence analysis not only for assessing agreement among strategies but also for informing multi-criteria selection frameworks that balance short-term phenotypic gains with long-term genetic progress.

The predicted genetic gains indicate that the MMI and BI indexes were the most efficient in predicting genetic gains for both grain yield (GY) and ear weight (EW), with values surpassing those obtained with SHI. The superiority of the MMI index for GY (15.93%) and the close performance of BI (15.84%) suggest that these methods are more effective at combining multiple traits into a selection criterion that maximizes productivity. SHI, although traditionally robust, presented slightly lower gains (14.79%), reinforcing the notion that alternative indices can provide higher efficiency in specific contexts.

Inbreeding depression (ID) indicates a consistent reduction in deleterious additive gene effects and that the selected progenies are likely to suffer lower losses under inbreeding, a desirable outcome in pre-breeding programs. The ability to reduce inbreeding depression while improving grain yield highlights the potential of these selection strategies to simultaneously enhance performance and genetic stability. The inbreeding values results suggest that while these progenies from F_2 commercial hybrids exhibit strong additive effects for grain yield, they may be more sensitive to inbreeding, a key consideration in early recurrent selection cycles.

The total predicted selection gain (PG total), reflecting the aggregate improvement across traits, was numerically greater with the Smith–Hazel index (28.25%), followed by the BI (26.37%) and MMI (26.27%), indicating that all three indexes were robust for multi-trait selection in half-sib progenies of tropical maize. Although SHI showed a numerically greater composite score, MMI and BI provided more favorable responses for key breeding objectives such as grain yield and reduction in inbreeding depression.

The findings of the present study align with those of Velmurugan et al. [13], who demonstrated significant genetic variability in BC_2F_1 maize progenies derived from *Zea nicaraguensis* crosses, particularly for kernel traits such as 100-seed weight, color, and shape. Our results and those of Velmrugan et al. [13] converge on the importance of maintaining and reintroducing genetic diversity as a strategy for improving maize performance under variable environmental conditions, and both highlight the value of multivariate analyses, such as mixed-model predictions, for guiding selection decisions in early recurrent selection cycles.

The REML/BLUP methodology was applied to estimate genetic gains and assess genotypic values among maize half-sib progenies under a recurrent selection scheme [5]. The REML/BLUP approach yielded high accuracy estimates and predicted substantial selection gains: 24.07% for EL, 21.88% for GY, and 18.23% for ED. These results underscore the efficiency of mixed models in quantifying genetic progress in maize breeding populations.

These findings reinforce the utility of BLUP-based methodologies in the estimation of genetic values and selection gains in maize, as implemented in our study. The positive association between yield components and GY through traits like ear length supports the integration of multivariate selection strategies, such as the SHI, BIA, and MMI indices, which incorporate multiple agronomic traits into the decision-making process. The genetic architecture observed by [5] aligns with our own findings of divergence among selection indices and suggests that trait-specific indirect selection paths may enhance the efficiency of breeding programs, particularly when the coincidence among selection methods is low.

The outcomes of our study, particularly the effectiveness of BLUP-based selection strategies in generating additive genetic gains and enhancing grain yield performance, are consistent with simulation-based insights presented by Vieira et al. [9]. Their review highlights that selection strategies incorporating genomic prediction tools—such as BLUP and RR-BLUP—are especially efficient when applied to traits with moderate to high heritability, as is typical for yield-related components in maize. In this study, the application of BLUP (PopBLUP_GY) resulted in superior grain yield compared to the open-pollinated control, validating simulation findings that emphasize the robustness of BLUP in handling polygenic traits across early recurrent selection cycles. This underscores the value of integrating both predictive models and cyclic selection schemes to simultaneously optimize short-term gains and long-term breeding potential in tropical maize.

The REML/BLUP framework has proven particularly effective in this context, providing precise estimates of additive genetic effects and enabling the identification of genotypes that combine high yield potential with resilience to inbreeding. As Kick and Washburn [18] demonstrated, predictive models that incorporate genotypic data can outperform phenotypic selection alone, particularly in unbalanced datasets or when dealing with complex traits.

In conclusion, the REML/BLUP-based evaluation provides a solid framework for informed selection decisions. Progenies with high additive genetic values and stable performance under inbreeding should be prioritized for hybrid development and population improvement, while lower-performing progenies may be strategically discarded or redirected to secondary breeding objectives.

The significance of the differences among new synthetic populations confirms the effectiveness of the selection and recombination processes applied to these new populations. Notably, PopMMI and PopBLUP_GY stood out as the top-performing groups, evidencing their potential for use in future hybrid combinations or for advancement as synthetic cultivars. These results reaffirm that the selection and recombination processes were effective in generating competitive and genetically diverse breeding materials.

As reported by Duarte and Sentelhas [20], the comparison with commercial hybrids is essential for determining the feasibility of replacing or complementing conventional materials with newly developed populations. In this case, although commercial hybrids still perform better in absolute terms, synthetic maize populations, especially those derived from the MMI and BLUP-based strategies, demonstrated yield levels that surpass traditional open-pollinated varieties and approach hybrid standards.

The genetic gains observed across progeny generations in our study are consistent with findings reported in a five-cycle intrapopulation recurrent selection program in half-sib progenies of fresh corn conducted in the southwest region of Goiás, Brazil. That study demonstrated the persistence of genetic variability and cumulative progress in key agronomic traits over multiple selection cycles, with heritability estimates exceeding 70% for most yield-related components, including ear diameter, marketable ear yield, and ear length [14]. Both studies underscore the efficacy of recurrent selection schemes in maintaining selection potential over time, even in populations derived from diverse genetic backgrounds, and highlight the utility of half-sib progeny structures for early-stage recurrent selection cycles and from F_2 commercial hybrids. These parallels reinforce the strategic value of population improvement via intrapopulation recurrent selection, not only for specialty markets like fresh corn but also for broad-based grain yield improvement.

The observed results reinforce that the recombination of half-sib progenies, when guided by structured strategies such as REML/BLUP-based selection, leads to the formation of differentiated and agronomically competitive populations. Collectively, these findings support the continued use of the RV-02 germplasm base as a cornerstone for recurrent selection programs focused on improving tropical maize in terms of yield, resilience, and genetic progress across cycles and environments.

In summary, the data support the conclusion that genetic gains were successfully achieved, particularly in PopMMI and PopBLUP_GY, which surpassed or closely approached commercial check performance in yield-related traits. These results validate the effectiveness of selection strategies applied and demonstrate the potential of these populations for advancing synthetic variety development or serving as sources for inbred line extraction.

When comparing the relative positions of the synthetic populations based on predicted and realized grain yield (GY), clear differences emerge that highlight the limitations of the predictive models. PopMMI achieved the highest realized yield (6300 kg ha^{-1}), although its predicted value (4590 kg ha^{-1}) placed it close to PopBIA and PopSHI. This underestimation suggests that the index did not fully capture favorable non-additive effects or environmental contributions expressed in the field.

PopBLUP_GY presented the most striking shift, being ranked lowest in predicted yield (4025 kg ha^{-1}) but outperforming PopBIA and PopSHI in realized yield (5620 kg ha^{-1}). This discrepancy indicates that the BLUP-based model underestimated its genetic potential, likely due to unaccounted dominance and epistatic effects, which are not fully represented in additive genetic predictions, probably by early genetic cycles from F_2 populations. Nevertheless, even though the predicted values were inferior to those obtained with phenotypic indices, the realized yield clearly demonstrates the efficiency of the BLUP-based method in selecting superior progenies with higher additive values. This outcome reinforces the strength of BLUP as a selection strategy, since its primary goal is not necessarily to maximize short-term prediction accuracy, but to identify and accumulate favorable additive effects that ensure sustained genetic progress across recurrent selection cycles.

Finally, the use of F_2 populations derived from commercial hybrids to synthesize base populations for recurrent selection cycles in maize breeding can be considered a valuable strategy, as these populations harbor a high frequency of favorable alleles that have already been combined through intensive selection in commercial programs. This allows the synthesized populations to start from a high-performing genetic pool, thereby accelerating initial genetic gains. However, such populations may also display increased homozygosity and greater sensitivity to inbreeding, given that commercial hybrids are derived from specific combinations of inbred lines. In this context, the adoption of half-sib progeny selection schemes is technically advisable, since they reduce the rate of inbreeding compared with full-sib or inbred progenies, while maintaining sufficient genetic variability for future progress. Therefore, while F_2 populations from hybrids can provide an efficient starting point for recurrent selection, their sustainable use requires strategies that balance the exploitation of favorable alleles with the preservation of long-term genetic diversity.

4. Materials and Methods

The study was carried out within a long-term maize breeding program in Southwest Goiás, Brazil, from 2016 to 2020 (Figure 4). In this region, the first season is sown in October of the previous year and harvested in February of the following year, while the second season is sown in February and harvested at the end of August. In 2016 (second season), the RV-02 population was planted in isolated plots to evaluate 182 half-sib families (HSP) across three locations, as described by Chavaglia (2016) [21]. In 2017 (second season), 32 superior HSP were recombined with 9 HSP previously identified for foliar disease resistance. In 2018 (first season), the recombination of 41 HSP generated a new population from which 194 HSP were selected. In 2018 (second season), 194 progenies were self-pollinated and obtained 194 S1 progenies. In 2019 (second season), these 194 HSP, together with two checks and their respective S1 progenies, were tested in field trials. In 2020 (first season), 38 selected HSP were recombined, resulting in 5 populations that were evaluated in 2020 (second season). These seven sown dates, with a chronological sequence of recombination, evaluation, and selection, ensured consistent replication across years, captured environmental variability, and established a robust framework for assessing the efficiency of the selection methods.

4.1. Recurrent Selection Cycle 0

The RV-02 Cycle 2 maize population was developed through a structured breeding and selection program over 15 years, serving as the genetic basis for the present study.

The original RV-02 Cycle 0 maize population originated from the work of Cárdenas [22], who collected germplasm from commercial maize fields in Southwest Goiás, Brazil, and established 14 populations (HG1 to HG14) from F_2 commercial hybrids (Figure 5).

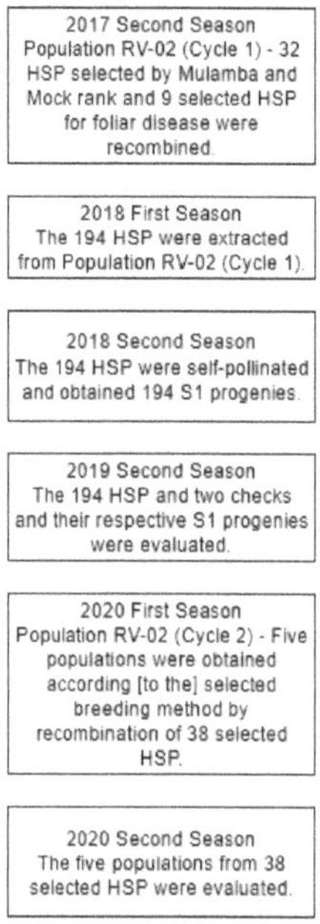

Figure 4. Flowchart illustrating the recurrent selection cycles of the RV-02 tropical maize population across seven sown dates.

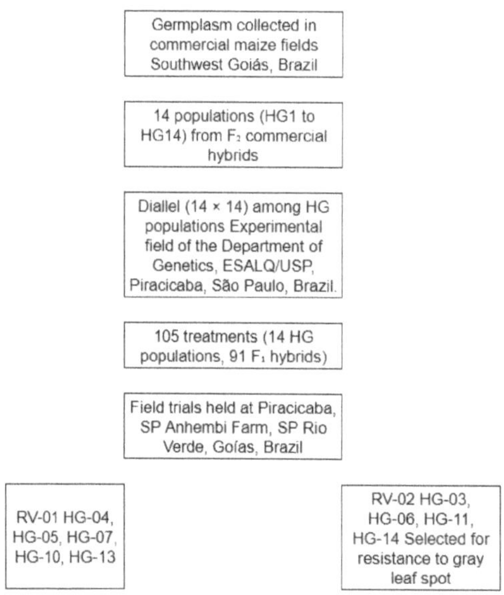

Figure 5. Flowchart of the initial development and selection process of RV-02 synthetic maize population.

A full diallel (14 × 14) among 14 HG populations was conducted in January 2001 at the experimental field of the Department of Genetics, "Luiz de Queiroz" College of Agriculture (ESALQ/USP), in Piracicaba, São Paulo, Brazil (Figure 5). A total of 91 full-sib progenies were generated through manual crosses between adjacent population rows. Fifty to sixty ears per pair of rows were pollinated. Field trials with 14 HG populations and 91 full-sib progenies were conducted at Piracicaba (SP), Anhembi Farm (SP), and Rio Verde (GO), Brazil (Figure 5).

Two synthetic populations were formed: RV-01 from the recombination of HG-04, HG-05, HG-07, HG-10, HG-13, and RV-02 from HG-03, HG-06, HG-09, HG-11, HG-14 (Figure 5). The RV-02 synthetic population was selected for resistance to gray leaf spot (*Cercospora zeae-maydis*).

4.2. Recurrent Selection Cycle 1

The first cycle of recurrent selection of base population RV-02 (C1) was obtained under selection of 182 half-sib progenies (HSP) in three locations in the Southwest of Goiás, Brazil, in 2016, in the second season. These progenies were initially evaluated by Chavaglia [21] for grain yield and foliar diseases, including southern rust (*Puccinia polysora*), gray leaf spot (Pantoea ananatis), northern leaf blight (*Exserohilum turcicum*), cercospora leaf spot (*Cercospora zeae-maydis*), and maize bushy stunt (*Spiroplasma kunkelii*). In 2017, in the second season, 32 half-sib progenies were selected using the Mulamba and Mock [23] rank summation index, assigning a weight of 4 to grain yield and a weight of 1 to each disease variable. An additional 9 half-sib progenies were selected specifically for foliar disease resistance, resulting in 41 selected progenies. These progenies were recombined using the Irish method, in which a bulked pollen mixture from the progenies was used to pollinate each emasculated progeny. It originated from the Population RV-02 (Cycle 1).

In 2018, in the first season, 194 HSP were extracted from Population RV-02 (Cycle 1), and in the second season, the 194 HSP were self-pollinated and obtained 194 S1 progenies.

4.3. Recurrent Selection Cycle 2

The 194 HSP were evaluated in the 2019 second season using a 14 × 14 triple lattice design with two checks: a single-cross hybrid (DKB 290) and an open-pollinated variety AL Bandeirante. In parallel, the S1 progenies were grown in a separate field to evaluate inbreeding depression. To control competition effects, S1 rows were flanked by S0 base population rows, with a configuration ensuring uniform growth conditions.

The experiments with 194 HSP and 194 S1 progenies were conducted at the experimental field of the Federal University of Jataí, Jataí, Goiás, Brazil (17°53′ S, 51°43′ W, 780 m altitude, average annual precipitation ~1800 mm).

The plots consisted of a single 4-m row with 20 plants, spaced 0.90 m between rows and 0.20 m between plants. The following traits were evaluated: male flowering (MF), female flowering (FF), ear height (EH), plant height (PH), number of ears (NE), ear length (EL), ear diameter (ED), ear weight (EW), grain yield (GY), grain moisture (GM), and inbreeding depression (ID). The ID was calculated as:

$$ID = 100 \times ((m_0 - m_1)/m_0), \quad (1)$$

where

m_0 = mean of S_0, and

m_1 = mean of S_1.

The second cycle of recurrent selection of Population RV-02 (C2) was obtained after recombination of 38 selected S1 progenies selected from 194 half-sib progenies in the 2020 first season (Figure 4). The 38 S1 selected progenies of each selection strategy in recurrent selection programs were recombined to generate five new populations: PopBIA (Base Index), PopSHI (Smith–Hazel Index), PopMMI (Mulamba–Mock Index), PopBLUP_GY (BLUP for grain yield), and PopBLUP_ID (BLUP for inbreeding depression).

4.4. Selection Strategies and Genetic Gain Prediction

Twenty percent of the best-performing progenies (38 out of 194 HSP) were selected based on five selection strategies: (i) phenotypic values using selection indexes Base Index, Smith and Hazel Index, and Rank Summation Index, and (ii) genotypic values (BLUPs) for grain yield and inbreeding depression. Selection indexes assigned weights of 2 (increase) for grain yield and 1 (decrease) for inbreeding depression. Genotypic selection considered these traits independently.

S1 progenies were used to calculate inbreeding depression and recombination, which were included as traits in all selection procedures. Selection indexes included the Base Index [24], the classical Smith [25] and Hazel [26] index (SHI), and the Mulamba and Mock [23] rank summation index (MMI):

$$\text{Base Index (BI): IBIj} = \textstyle\sum \text{j}\ x_j\ a_j, \quad (2)$$

where x_j are standardized trait means (per progeny) and a_j are predetermined economic/subject-matter weights.

$$\text{Smith–Hazel (SHI): ISHI} = b' \times I \text{ with } b = P^{-1}\ Ga, \quad (3)$$

where P and G are phenotypic and genetic (co)variance matrices, a is the vector of economic weights. I is the vector (t × 1) of values of genotype j.

$$\text{Mulamba \& Mock rank-sum (MMI): IMMI} = \textstyle\sum \text{j}\ r_{ij}\ w_i, \quad (4)$$

where rij is the rank of genotype j for trait i, and wi denotes the weights. The lower ranks for inbreeding depression and higher ranks for grain yield were favored. Weights were set to 2 for grain yield and 1 for inbreeding depression (other traits' weights = 0) in the phenotypic selection scenario. For MMI, lower ranks for inbreeding depression and higher ranks for grain yield were favored.

Genotypic evaluations used mixed models (REML/BLUP), with the following linear model:

$$Y = Xr + Zg + Wp + Ti + \varepsilon, \quad (5)$$

where Y is the phenotypic vector;

Xr = random repetition effects;

Zg = random genotypic effects;

Wp = random plot effects;

Ti = fixed ancestry matrix; and

ε = random residuals.

The fixed ancestry matrix coded the pedigree/background groups that formed the RV-02 population from F_2 commercial-hybrid sources (HG03...HG14 as per the RV-02 synthesis described by Cárdenas [22]). These fixed effects were included to absorb systematic ancestry structure during REML/BLUP estimation.

Predicted gains were computed from index-based responses (BI/SHI/MMI) using selected-top-20% progenies and, for BLUP-based strategies, from the mean of selected BLUPs relative to the base mean, expressed as percent gain per trait; selection considered yield (weight = 2) and inbreeding depression (weight = 1) together in the index scenario, and separately in BLUP scenarios.

Realized gains were estimated after recombining the selected sets into five derived populations (PopBIA, PopSHI, PopMMI, PopBLUP_GY, PopBLUP_ID) and evaluating them in RCBD. The gains were computed as the difference (or % difference) between recombined population means and the reference/base.

The agreement among selection strategies over the universe of genotypes was quantified. For each strategy s and selection intensity k, the top-k selection set $S_{s,k}$ was defined with cardinality k.

$$S_{s,k} \subseteq U, \ |S_{s,k}| = k$$

Pairwise coincidence (overlap) between strategies p and q at intensity k was computed as:

$$CI_{p,q}(k) = |S_{p,k} \cap S_{q,k}| / k$$

When selected sets had different sizes, the normalized version was reported:

$$CI_{p,q}(k) = |S_{p,k} \cap S_{q,k}| / \min\{|S_{p,k}|, |S_{q,k}|\}$$

As a sensitivity analysis, we also computed the Jaccard similarity:

$$J_{p,q}(k) = |S_{p,k} \cap S_{q,k}| / |S_{p,k} \cup S_{q,k}|$$

To summarize stability across all M strategies, we used the mean pairwise coincidence:

$$\overline{CI}(k) = 2/[M(M-1)] \cdot \Sigma_{p<q} CI_{p,q}(k)$$

The top 10 until the top 40 corresponded to the highest-ranking subsets of progenies selected by each method. In practice, all progenies were first ranked according to their respective criteria, predicted genetic values for BLUP-based strategies or index scores for phenotypic indexes. From these ordered lists, the top 10 to the top 40 progenies were extracted, and pairwise coincidence indices were subsequently calculated to assess the degree of overlap among methods.

Estimations were performed using the lme4 package in R [27].

4.5. Recombination and Evaluation of Recombined Progenies

The five populations comprised the population RV-02 (C2), which was evaluated in 2020, in the second season. Field trials used a randomized complete block design (RCBD) with three replications and seven treatments. The plots with 4 lines, spaced by 0.9 m, were used and the same traits were measured in the HSP and S1 progenies of the RV-02 (C2) populations.

5. Conclusions

The evaluation of five selection strategies in maize recurrent selection showed that both phenotypic indexes (BIA, SHI, and MMI) and BLUP-based approaches (BLUP_GY and BLUP_ID) were effective in identifying superior progenies. Some progenies were consistently selected across all methods, confirming their reliability for recombination. Although predicted and realized values differed, particularly for BLUP_GY, the overall results indicate that phenotypic and BLUP-based strategies are complementary: phenotypic indexes provided higher predicted short-term gains, while BLUP approaches proved efficient in capturing additive effects and reducing inbreeding. Therefore, integrating both methods can strengthen recurrent selection by combining immediate realized gains with sustainable genetic progress for tropical maize improvement.

Author Contributions: Conceptualization, A.M.d.C.B.d.S., M.P.M.R. and G.V.M.; methodology, A.M.d.C.B.d.S. and A.J.C.-F.; software, A.J.C.-F.; validation, A.M.d.C.B.d.S., M.P.M.R. and E.F.d.R.; formal analysis, A.M.d.C.B.d.S. and A.J.C.-F.; investigation, A.M.d.C.B.d.S. and M.P.M.R.; resources, M.P.M.R. and E.F.d.R.; data curation, A.J.C.-F. and A.M.d.C.B.d.S.; writing—original draft preparation, A.M.d.C.B.d.S. and M.P.M.R.; writing—review and editing, G.V.M. and E.F.d.R.; visualization,

A.J.C.-F.; supervision, G.V.M. and E.F.d.R.; project administration, M.P.M.R. and E.F.d.R.; funding acquisition, G.V.M. and E.F.d.R. All authors have read and agreed to the published version of the manuscript.

Funding: This research received no external funding.

Data Availability Statement: The data presented in this study are available on request from the corresponding author.

Acknowledgments: The authors wish to thank the Universidade Federal de Goiás, Universidade Federal de Jataí, and Universidade Tecnológica Federal do Paraná.

Conflicts of Interest: The authors declare no conflicts of interest.

References

1. Freitas Júnior, S.P.; Amaral Júnior, A.T.; Rangel, R.M.; Viana, A.P. Prediction of genetic gains in the UNB-2U popcorn population under recurrent selection using different selection indices. *Semin. Ciências Agrárias* **2009**, *30*, 803–814. [CrossRef]
2. Fritsche Neto, R.; Miranda, G.V.; De Souza, L.V.; De Lima, R.O.; De Oliveira, L.R. Prediction of genotypic values of maize hybrids in unbalanced experiments. *Crop Breed. Appl. Biotechnol.* **2010**, *10*, 32–39. [CrossRef]
3. Olivoto, T.; Souza, V.Q.; Nardino, M.; Carvalho, I.R.; Ferrari, M.; Pelegrin, A.J.; Szareski, V.J.; Schmidt, D. REML/BLUP and sequential path analysis in estimating genotypic values and interrelationships among simple maize grain yield-related traits. *Genet. Mol. Res.* **2017**, *16*, 1–19. [CrossRef]
4. Viana, J.M.S.; Dias, K.O.G.; Silva, J.P.A. Comparative analysis of pedigree-based BLUP and phenotypic mass selection for developing elite inbred lines, based on field and simulated data. *Agronomy* **2022**, *12*, 2560. [CrossRef]
5. Mascarenhas, S.N.A.P.; Amaral Júnior, A.T.; Gonçalves, G.M.; Vivas, M.; Oliveira, R.A.; Carvalho, B.L. Prediction of genotypic values via REML/BLUP and path analysis in maize half-sib progenies. *Cad. Pedagógico* **2024**, *21*, 1–15. [CrossRef]
6. Tarekegne, A.; Wegary, D.; Cairns, J.E.; Zaman-Allah, M.; Beyene, Y.; Negera, D.; Teklewold, A.; Tesfaye, K.; Jumbo, M.; Das, B.; et al. Genetic gains in early maturing maize hybrids developed by the International Maize and Wheat Improvement Center in Southern Africa during 2000–2018. *Front. Plant Sci.* **2024**, *14*, 1321308. [CrossRef]
7. Entringer, G.C.; Vettorazzi, J.C.F.; Santos, E.A.; Oliveira, R.B.; Scapim, C.A. Genetic gain estimates and selection of S1 progenies based on selection indices and REML/BLUP in super sweet corn. *Aust. J. Crop Sci.* **2016**, *10*, 330–337. [CrossRef]
8. Pedrozo, C.Á.; Barbosa, M.H.P.; Resende, M.D.V.; Peternelli, L.A.; Costa, P.M.A.; Silva, F.L. Efficiency of selection indexes using the REML/BLUP procedure in sugarcane breeding. *Sci. Agrar.* **2008**, *10*, 31–36. [CrossRef]
9. Simeão, R.M.; Resende, M.D.V.; Alves, R.S.; Pessoa-Filho, M.; Azevedo, A.L.S.; Jones, C.S.; Pereira, J.F.; Machado, J.C. Genomic Selection in Tropical Forage Grasses: Current Status and Future Applications. *Front. Plant Sci.* **2021**, *12*, 665195. [CrossRef] [PubMed]
10. Melchinger, A.E.; Frisch, M. Genomic prediction in hybrid breeding: II. Reciprocal recurrent genomic selection with full-sib and half-sib progenies. *Theor. Appl. Genet.* **2023**, *136*, 203. [CrossRef]
11. Melchinger, A.E.; Fernando, R.; Melchinger, A.J.; Frisch, M. Optimizing selection based on BLUPs or BLUEs in multiple sets of genotypes differing in their population parameters. *Theor. Appl. Genet.* **2024**, *137*, 104. [CrossRef]
12. Farid, M.; Azrai, M.; Nur, A.; Syahruddin, K.; Efendi, R.; Suwarno, W.B.; Andayani, N.N.; Priyanto, S.B.; Mukminati, M.; Anshori, M.F. Diversity analysis of convergent-derived breeding maize inbred lines using SSR markers. *Int. J. Agric. Biosci.* **2025**, *14*, 596–605. [CrossRef]
13. Senthilkumar, V.; Garkoti, P.; Varalakshmi, S.; Naresh, T.; Singh, N.K. Kernel trait and genetic variability analysis in BC_2F_1 maize lines derived from *Zea nicaraguensis* crosses. *J. Adv. Biol. Biotechnol.* **2024**, *27*, 268–279. [CrossRef]
14. Cintra, P.H.N.; Resende, C.L.P.; Damaso, L.F.; Carvalho, D.D.C.; Silva, F.C.; Rodrigues, F. Five cycles of intrapopulation recurrent selection in half-sib progenies of fresh corn. *Rev. Caatinga* **2023**, *36*, 723–730. [CrossRef]
15. Kist, V.; Bernardi Ogliari, J.; De Miranda Filho, J.B.; Souza, C.L., Jr. Genetic potential of a maize population from Southern Brazil for the modified convergent–divergent selection scheme. *Euphytica* **2010**, *176*, 25–36. [CrossRef]
16. Pedro, C.; Marçola, M.A.; Charimba, A.M.; Queiroz, L.G.C.; Souza, J.C. Genetic potential of maize full-sib progenies subjected to a reciprocal recurrent selection. *Pesqui. Agropecuária Bras.* **2023**, *58*, e03134. [CrossRef]
17. Garcia, A.A.F.; Frisch, M.; Weng, Y.; Melchinger, A.E. Heterosis and hybrid breeding. *Theor. Appl. Genet.* **2025**, *138*, 69. [CrossRef]
18. Kick, D.R.; Washburn, J.D. Ensemble of BLUP, machine learning, and deep learning models predict maize yield better than each model alone. *bioRxiv* **2023**. [CrossRef]
19. Berilli, A.P.C.G.; Eweira, M.G.; Trindade, R.S.; Costa, F.R.; Cunha, K.S. Response to the selection in the 11th cycle of reciprocal recurrent selection among fullsib progenies of maize. *Acta Sci. Agron.* **2013**, *35*, 435–441. [CrossRef]

20. Duarte, Y.C.N.; Sentelhas, P.C. Intercomparison and performance of maize crop models and their ensemble for yield simulations in Brazil. *Int. J. Plant Prod.* **2020**, *14*, 453–467. [CrossRef]
21. Chavaglia, A.C. Productive Potential, Genetic Variability, and Inbreeding Depression in a Population Derived from Commercial Maize Hybrids. Master's Thesis, Universidade Federal de Goiás, Jataí, Brazil, 2017.
22. Cárdenas, F.E.N. Valor Genético de Populações de Milho Adaptadas Para a Região Sudoeste de Goiás. Ph.D. Thesis, Escola Superior de Agricultura "Luiz de Queiroz"—ESALQ, Universidade de São Paulo—USP, Piracicaba, Brazil, 2005; 169p.
23. Mulamba, N.N.; Mock, J.J. Improvement of yield potential of the Eto Blanco maize (*Zea mays* L.) population by breeding for plant traits. *Egypt. J. Genet. Cytol.* **1978**, *7*, 40–51.
24. Williams, J.S. The evaluation of a selection index. *Biometrics* **1962**, *18*, 375–393. [CrossRef]
25. Smith, H.F. A discriminant function for plant selection. *Ann. Eugen.* **1936**, *7*, 240–250. [CrossRef]
26. Hazel, H.N. The genetic basis for constructing selection indexes. *Genetics* **1943**, *28*, 476–490. [CrossRef] [PubMed]
27. R Core Team. *R: A Language and Environment for Statistical Computing*; R Foundation for Statistical Computing: Vienna, Austria, 2018. Available online: https://www.r-project.org/ (accessed on 7 July 2025).

plants

MDPI

Article

The Identification of a Single-Base Mutation in the Maize *Dwarf 1* Gene Responsible for Reduced Plant Height in the Mutant 16N125

Ping Wang [1], Bingbing Liang [1], Zhengjun Li [2], Huaiyu Dong [1], Lixia Zhang [2,*] and Xiaochun Lu [2,*]

[1] Institute of Plant Protection, Liaoning Academy of Agricultural Sciences, Shenyang 110161, China; pingw-556@163.com
[2] Institute of Sorghum, Liaoning Academy of Agricultural Sciences, Shenyang 110161, China
* Correspondence: zhanglx993@sohu.com (L.Z.); luxiaocun2000@126.com (X.L.)

Abstract: Maize (*Zea mays* L.) is a globally vital crop for food, feed, and biofuel production, with plant height (PH) being a key agronomic trait that significantly influences yield, lodging resistance, and stress tolerance. This study identified a single-base mutation in the *D1* (*Dwarf 1*) gene responsible for the dwarf phenotype in the maize mutant 16N125. Through genetic analysis and fine mapping, the candidate region was localized to chromosome 3, narrowing it down to an interval containing three genes. Sequencing revealed a non-synonymous mutation in *D1*, which encodes a gibberellin 3-beta-dioxygenase, leading to amino acid substitutions at positions 61 and 123. Genetic analysis of F2 populations confirmed that the mutation at position 61 was responsible for the dwarf trait. Furthermore, the mutation was detected in several Chinese inbred lines, indicating its potential role in dwarfing under specific conditions. These findings provide critical insights into the genetic mechanisms regulating maize plant height, offering valuable information for breeding programs focused on improving crop architecture and yield to address the challenges of global food security and climate change.

Keywords: maize; plant heigh; *D1* gene; dwarf mutant; genetic mapping; breeding

1. Introduction

Maize (*Zea mays* L.) is one of the most important global crops serving as a staple food, feed, and biofuel source. It is a cornerstone of global agriculture, providing essential nutrients and energy for both humans and livestock. Over the past century, maize production has increased eightfold, largely due to genetic improvements, advanced agricultural practices, and the development of high-yielding hybrids [1–3]. However, with the growing global population, which is projected to reach nearly 10 billion by 2050, and the challenges posed by climate change, there is an urgent need to further enhance maize productivity, resilience, and adaptability [4,5]. Among the various agronomic traits that influence maize yield and performance, plant height (PH) stands out as a critical factor [6,7]. Plant height serves as a crucial morphological trait that profoundly influences stress tolerance in plants. Taller plants exhibit superior light capture efficiency, particularly in dense canopies, thereby enhancing their photosynthetic capacity under light-limiting conditions. This architectural advantage also extends to thermal regulation, as their vertical profile enables exposure to varied microclimates that facilitate heat dissipation during high-temperature stress. Furthermore, shorter plants enhance lodging resistance through improved stem strength and a lower center of gravity, reducing yield losses under mechanical stress [6–10].

Plant height is a complex trait influenced by a combination of genetic, environmental, and hormonal factors. It is closely associated with other important traits such as ear height, internode length, and node number, which collectively contribute to biomass production and yield potential [11,12]. In energy maize, taller plants correlate strongly with a higher biomass, ideal for bioenergy sources like cellulosic ethanol. However, in grain maize, excessive height risks lodging, reducing yield and harvest efficiency. Breeding must thus balance biomass production and structural strength based on end-use goals [13].

The importance of plant height in crop improvement was highlighted during the Green Revolution of the mid-20th century, when the discovery and utilization of semi-dwarf genes (*rht1* in wheat and *sd1* in rice) revolutionized cereal production. These semi-dwarf varieties exhibited improved lodging resistance, increased yield, and enhanced responsiveness to fertilizers, leading to significant gains in global food production [14–16]. In maize, however, the genetic mechanisms controlling plant height are more complex, and identifying the genes that regulate this trait has proven to be more challenging. Nevertheless, the potential benefits of optimizing plant height in maize are immense, as it could lead to the development of high-yielding, stress-resistant varieties that are better suited to the changing climatic conditions and the increasing demand for food and bioenergy [17].

To date, over 50 genes have been implicated in the regulation of maize plant height, many of which are involved in the biosynthesis, transport, and signaling pathways of key plant hormones such as gibberellins (GAs), abscisic acid, and brassinosteroids [18–21]. Gibberellins, in particular, play a central role in regulating plant growth and development, including stem elongation, leaf expansion, and flowering. Mutations in genes involved in gibberellin biosynthesis or signaling often result in dwarf phenotypes, as seen in the well-studied *dwarf1* (*D1*) gene, which encodes a gibberellin 3-beta-dioxygenase [22–25]. Other genes, such as *ZmBr2* and *ZmGA2ox3*, have also been shown to influence plant height by modulating hormone levels and cell elongation [19,26].

Despite these advances, the genetic mechanisms underlying the natural variation in maize plant height remain poorly understood, primarily due to the complexity of the trait and the limited number of successfully cloned genes. Traditional linkage mapping and quantitative trait locus (QTL) analysis have identified numerous genomic regions associated with plant height, but the resolution of these methods is often insufficient for precise gene identification [27,28]. Moreover, the interaction between multiple genes and environmental factors further complicates the genetic dissection of plant height. In recent years, however, advances in genomics and sequencing technologies have provided new tools for studying complex traits in crops. One such tool is bulked segregant analysis combined with next-generation sequencing (BSA-seq), which has emerged as a powerful method for rapidly identifying candidate genes associated with specific traits [29–32]. This approach significantly reduces the number of samples required for genotyping and focuses on individuals with extreme phenotypes, making it particularly suitable for studying complex traits such as plant height.

In this study, we employed BSA-seq to identify a novel mutation in the *D1* gene responsible for the dwarf phenotype in the maize mutant 16N125. The *D1* gene, also known as ZmGA3ox2, is a key regulator of gibberellin biosynthesis and has been associated with several dwarfing alleles in maize [21–25]. By combining BSA-seq with fine mapping and genetic analysis, we were able to narrow down the candidate region to an interval on chromosome 3 and identify a non-synonymous mutation in the *D1* gene that leads to amino acid substitutions at positions 61 and 123. Genetic analysis of F2 populations confirmed that the mutation at position 61 was responsible for the dwarf trait also found in several Chinese inbred lines, suggesting its potential role in dwarfing under specific conditions.

We aimed to identify the gene responsible for the dwarf phenotype in the maize mutant 16N125 through genetic analysis and BSA-seq. We crossed 16N125 with different maize varieties to investigate the inheritance pattern of the dwarf trait. BSA-seq and fine mapping were then employed to pinpoint the candidate gene. Furthermore, we examined the presence of the identified mutation in Chinese inbred lines to explore its potential role in maize breeding. This research is expected to contribute to a better understanding of the genetic mechanisms underlying maize plant height and provide practical guidance for future maize breeding programs. By uncovering the genetic basis of plant height in maize, we can develop more efficient breeding strategies to meet the growing demands for food, feed, and fuel, while also enhancing the resilience of maize crops in the face of a changing climate.

2. Results

2.1. Phenotyping and Genetic Analysis of Dwarf Maize Plants

In the summer of 2015, a mutant plant exhibiting extremely reduced stature was identified in an experimental field planted with B73, representing a naturally occurring mutation. If the naturally mutated gene can be mapped, it may represent a novel allele influencing maize plant height. Consequently, we retained this serendipitously obtained mutant and propagated it continuously for seven generations until the traits stabilized. This mutant was named 16N125 to identify the gene influencing plant height. We conducted a series of crosses between this mutant and multiple maize varieties and successfully obtained F1 generation seeds. In the following spring, these F1 seeds were sown in Shenyang city (42° N,123° E), and no significant changes in plant height or panicle length were observed, indicating a phenotype similar to that of the tall parental generation. Subsequently, after self-pollinating the F1 plants, the F2 generation was generated, and upon sowing the F2 seeds, clear trait segregation was evident. Some plants displayed a marked reduction in both plant height and panicle length. However, the segregation ratio of the F2 generation exhibited a discernible pattern solely when crossed with maize variety A101—a tall, inbred line selected and bred from America. Statistical analysis revealed that the dwarf maize plants had an average height of approximately 120 cm and the tall ones were 210 cm, confirming that the dwarf trait was completely recessive to the tall trait (Figure 1). Additionally, the numbers of tall and short plants are summarized in Table 1. According to the law of segregation, in F2 populations, the ratio of individuals with dominant traits to those with recessive homozygous traits should be 3:1. Among the 825 F2 individuals, 609 displayed a tall plant height while 216 showed a dwarf plant height. When subjected to the Chi-square test, the segregation ratio was found to be 3:1 ($\chi^2 = 0.62 < 3.84$, and the p value was greater than 0.05). The fact that $\chi^2 < 3.84$ indicates a significant correlation between the theoretical segregation ratio and the actual one. These results thus demonstrated that a single recessive gene should be responsible for controlling the dwarf trait.

Table 1. Phenotype analysis for tall/dwarf plant height in F2 populations.

Generation	Total Plants	Tall	Dwarf	[a] Expected Ratio	[b] χ^2	[c] p Value
F_2	825	609	216	3:1	0.62	0.43

[a] Expected ratio of dominant trait individuals–recessive trait individuals = 3:1. [b] χ^2: $\chi^2 < 3.84$ is considered as significantly correlated. [c] p value: p value > 0.05 indicates no statistically significant differences.

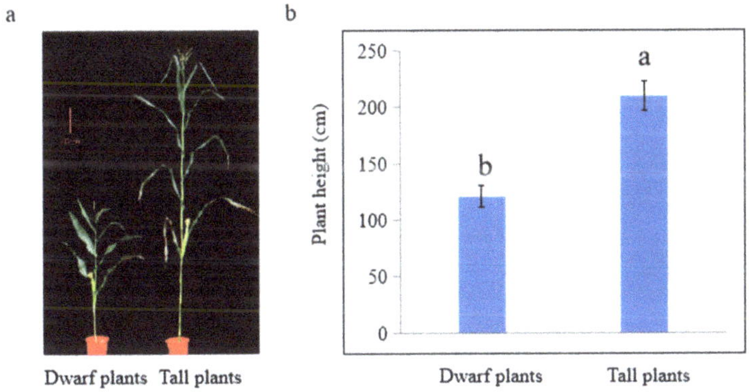

Figure 1. Plant height statistics of F2 generation plants. (**a**) Representative photographs of tall and dwarf plants in F2 generation; (**b**) statistical analysis of plant height. Values are means ± SD (n = 10). Columns marked with different letters (a–b) indicate significant differences, analyzed by SPSS software version 22.0 (Duncan's multiple range test, α = 0.05). Bar = 20 cm.

2.2. BSA-Seq Analysis and Mapping of Candidate Gene

To identify novel genes influencing plant height, we constructed a population for bulked segregant analysis sequencing (BSA-seq). Fifty plants exhibiting extreme phenotypes were selected to form either the extremely tall or dwarf pools (Table 2). Using the Illumina HiSeq platform, we obtained approximately 2.2 billion raw reads from sequencing the parental plants and extreme bulks (Table 2). Following stringent quality control procedures, including adapter trimming and low-quality read removal, we retained 330.2 billion high-quality clean bases.

Table 2. Sequencing statistics for BSA-seq samples.

Sample	Clean Reads	Clean Bases	GC (%)	Q20 (%)	Q30 (%)
A101	168,863,930	25,329,589,500	0.4631	0.9741	0.9296
16N125	172,965,252	25,944,787,800	0.4695	0.9699	0.9201
Dwarf	913,765,764	137,064,864,600	0.4653	0.9762	0.9337
Tall	945,640,034	141,846,005,100	0.4657	0.9732	0.9274
Sum	2,201,234,980	330,185,247,000	0.4659	0.9733	0.9277

The quality assessment demonstrated an excellent sequencing performance, with GC content ranging from 46.31% to 46.95%. The data quality metrics exceeded standard thresholds, exhibiting Q20 ≥ 96.99% and Q30 ≥ 92.01% (Table 2). These results confirm the high reliability of our sequencing data for subsequent genetic analyses.

Single-nucleotide polymorphism (SNP) detection and filtering were performed using the GATK software toolkit (version 4.1.8.1). To identify genomic regions associated with plant height, we applied the ΔSNP-index method to assess allele frequency differences in SNPs and InDels between the two extreme bulk pools. Association intervals were determined based on fitted ΔSNP-index values, with a 99% confidence threshold (red line, Figure 2). The significant regions are summarized in Table 3 and marked by arrows in Figure 2. The most robust association was detected within a 5.6 Mb interval (7,000,000–12,600,000 bp) on chromosome 3.

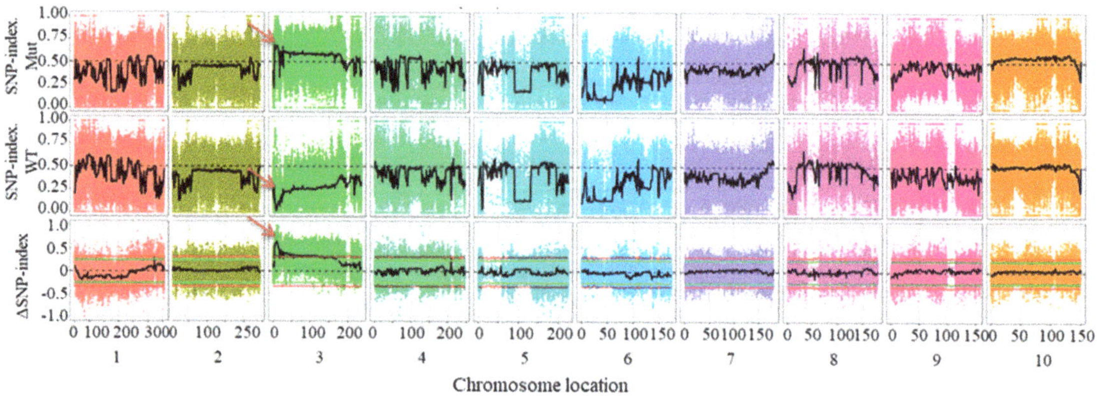

Figure 2. The genome-wide association mapping of plant height. The candidate SNPs and InDels when using the ΔSNP−index algorithm with a cutoff of ΔSNP−index > 0.5. The arrow indicates the region where the predicted candidate gene was located.

Table 3. Statistics of relevant area information.

Chromosome ID	Start	End	Size (Mb)
3	3,800,000	17,900,000	14.1
3	18,400,000	28,000,000	9.6
3	28,700,000	55,300,000	26.6
3	55,800,000	122,100,000	66.3
3	126,300,000	131,500,000	5.2
3	132,300,000	134,200,000	1.9
3	151,300,000	152,200,000	0.9
3	152,500,000	153,200,000	0.7
3	154,200,000	154,300,000	0.1

2.3. *Fine Mapping of Candidate Gene*

Based on the above BSA-seq results, 14 InDel markers were developed on chromosome 3, which covered a slightly larger region than predicted and displayed stable polymorphisms between the parental line and the F2 generation individuals whose parents were 16N125 and A101 (Figure 3). By using a recombinant-derived progeny testing strategy [33,34], we screened 752 F2 plants and identified 248 individuals recombinant between the markers InDel-1 and InDel-14, as exhibited in Figure 3.

Through successive [33,34] screenings, we identified 38 recombinants, narrowing the candidate region to the interval between markers InDel-2 and InDel-10. To further refine the location, we developed six additional InDel markers and screened for rare recombinants. The final two recombinants delimited the critical region to a 240 kb segment (9.57–9.81 Mbp) containing only three annotated genes (Figure 3, Table 4). Notably, *D1*, a strong candidate gene, was located within this interval.

Table 4. Gene description from NCBI of 3 genes in predictive region.

Gene Number	Annotation
Zm00001d039632	Putative polyol transporter 1, *Arabidopsis thaliana* (mouse-ear cress)
Zm00001d039633	Pleiotropic drug resistance protein 15, *Oryza sativa* subsp. *japonica* (rice)
Zm00001d039634	Gibberellin 3-beta-dioxygenase 2-2, *Triticum aestivum* (wheat)

Genes were blasted on NCBI website (https://www.ncbi.nlm.nih.gov/ (accessed on 4 December 2023)).

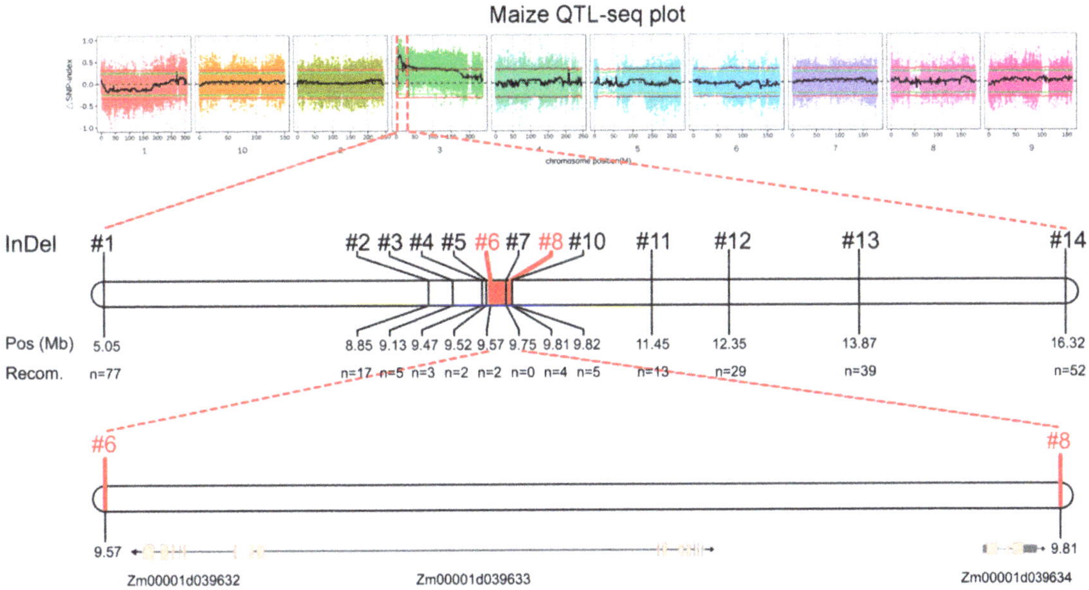

Figure 3. The fine mapping of the candidate gene, narrowing the candidate region through a recombinant-derived progeny testing strategy. The genetic linkage map of the candidate region of Chr.3 and the genes and diagnostic markers in the 11.27 Mb interval.

2.4. Identification of Candidate Gene Related to Plant Height in 16N125 Plants

To pinpoint the causal gene, we conducted Sanger sequencing for all three candidate genes in both parental lines. While nucleotide variations were present in all three genes, only *D1* exhibited a non-synonymous mutation in the amino acid sequence of 16N125 plants (Figure 4). For further analysis, we aligned and compared the coding sequences (CDSs; Figure 4a,b) and corresponding amino acid sequences (Figure 4c,d) of *D1* to illustrate the functional impact of this mutation. In contrast to B37, the amino acid sequence of 16N125 contained a substitution of alanine for proline at position 61 (Figure 4c) and asparagine for isoleucine at position 123 (Figure 4d), which was caused by the conversion of "C" to "G" at position 181 (Figure 4a) and "T" to "A" at position 368 (Figure 4b) in the CDS, respectively. Interestingly, the two substitutions in 16N125 were located at the sites corresponding to *d1-4*, a previously reported allele of *d1* that contains a large sequence deletion. The mutation sites in 16N125 plants were found within the regions of these missing fragments. Therefore, we concluded that the mutation in the *D1* gene might be responsible for the dwarfing phenotype observed in the 16N125 plants. To determine which of the two mutation sites played a role in the dwarfing of the 16N125 plants, we examined the sequence of the *D1* gene using Sanger sequencing in some randomly selected extremely tall and extremely dwarf plants from the F2 generation, and the results are listed in the Table S1. Among the eight dwarfing progenies, seven exhibited a homozygous G at position 181, identical to that of 16N125. In contrast, the tall progenies either possessed a C at position 181, similar to A101, or were heterozygous with C/G at this position. The phenotype resulting from this gene mutation aligns with the characteristics of a recessive gene. Notably, while the dwarf plants showed consistency with 16N125 at position 368, no discernible pattern emerged in the gene sequence of the tall offsprings. Interestingly, five tall plants were homozygous for allele A at this locus, which is consistent with the genotype of the dwarf parent 16N125. In summary, the mutation at position 61, rather than the other one, plays a regulatory role in maize plant height.

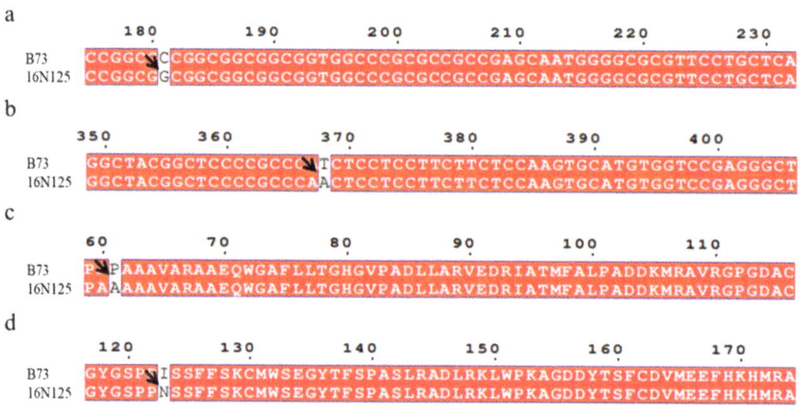

Figure 4. CDS and amino acid sequence analysis of *D1* genes from different cultivars. CDSs (**a**,**b**) and amino acid (**c**,**d**) sequences of *D1* arranged to show variations. Multiple sequence alignment performed by espript 3.0 online tool (https://espript.ibcp.fr/ESPript/cgi-bin/ESPript.cgi (accessed on 4 December 2024)). Arrows point to locations of variations in sequences.

2.5. Investigation of D1 Mutant Genes in Widely Utilized Chinese Inbred Lines

To investigate the amino acid mutation sites of the *D1* gene in 16N125 plants within a natural population, we sowed 50 commonly available Chinese inbred lines, including B73, in 2023. We then analyzed the variation in the *D1* gene among these lines relative to B73 using Sanger sequencing. Additionally, plant height was measured at maturity. As shown in Figure 5, 18 inbred lines exhibited a significant reduction in plant height compared to B73. Notably, the amino acid sequences of the *D1* gene in Ke830, Kehai181, Chang3, and Sui8941 were identical to those of B73, suggesting that the reduced height in these four lines may be attributed to other dwarfing genes or factors (Supplementary File S1).

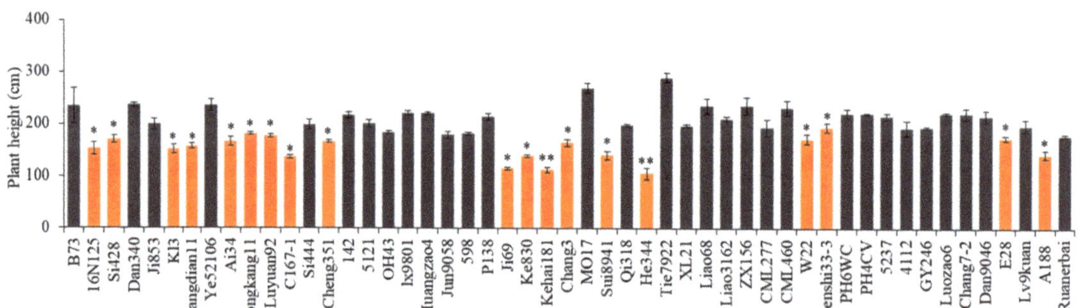

Figure 5. Plant height statistics of 50 commonly used inbred lines and 16N125. Plant height investigated ($n \geq 15$) in 2022. Asterisks indicate statistically significant differences compared with B73 using Student's *t*-test (*, $p < 0.05$; **, $p < 0.01$).

The results above indicate a substitution from proline to alanine at position 61 of the 16N125 amino acid sequence. Similarly, several inbred lines exhibiting reduced plant height also possessed alanine at this position, including Si428, Longkang11, Luyuan92, W22, E28, and A188. Additionally, glutamine was observed at position 61 in other lines, such as C167-1, Cheng351, Ji69, and He344 (Supplementary File S1). Therefore, 76.92% of the

13 inbred lines that exhibited a change in the *D1* amino acid sequence and a corresponding decrease in plant height demonstrated a mutation at amino acid position 61. The cause of their dwarfism remains unclear. Additionally, all inbred lines exhibiting a height more significant than that of B73 possessed proline at amino acid position 61, which is consistent with the residue found in B73. Based on these findings, it can be inferred that the 61st amino acid of the maize *D1* gene may influence plant height under specific conditions. However, the mutation at the 123rd amino acid position in the *D1* gene of 16N125 appears to be an isolated event, as no asparagine was detected at this position in other natural populations. Other mutations in the *D1* gene of various inbred lines compared to B73 are summarized in Supplementary File S1, but none were found to be significantly associated with changes in plant height. The variations in the *D1* gene for the remaining lines are detailed in Supplementary File S1. Notably, only three inbred lines—Kl3, Kangdian11, and Ai34—displayed a dwarfing phenotype despite retaining proline at position 61. The experiment was repeated three times, with similar results obtained.

3. Discussion

In the dynamic field of maize research, the exploration of genetic mechanisms underlying important traits has been significantly advanced by the advent of high-throughput sequencing technologies, with BSA-seq being a prominent example. In recent years, numerous studies have capitalized on BSA-seq to dissect the genetic architecture of various maize traits. Yu et al. (2024) employed BSA-seq to map chromosomal regions associated with low-temperature tolerance during maize germination [35]. Their research not only identified potential candidate genes but also provided valuable insights into the complex genetic network governing cold stress response in maize. Similarly, Zheng et al. (2023) combined BSA-seq and RNA-seq to identify genes associated with the visual stay-green phenotype during the maize maturation stage [36]. These studies not only showcase the power of BSA-seq but also lay a solid foundation for our investigation into the genetic basis of maize.

Plant height critically affects stress tolerance. Taller plants capture more light in dense canopies, improving photosynthesis under low light. Their height also aids heat dissipation by accessing cooler air layers. Conversely, shorter plants resist lodging better due to their stronger stems and lower center of gravity, protecting yield under mechanical stress [6–10]. In energy maize, taller plants mean more biomass for bioenergy. But in grain maize, too much height causes lodging, cutting yield. Breeding needs to balance biomass and strength for different uses [13].

Our study focused on identifying the gene responsible for the dwarf phenotype in the maize mutant 16N125. Through a comprehensive approach involving genetic analysis and BSA-seq, we successfully narrowed down the candidate gene region to chromosome 3. Subsequent fine-mapping and Sanger sequencing revealed a non-synonymous mutation in the *D1* gene, which encodes a gibberellin 3-beta-dioxygenase, as the cause of the dwarfing effect in 16N125. This finding aligns with the well-established role of the *D1* gene in regulating plant height through its involvement in the gibberellin biosynthesis pathway [22,23,25].

To date, four distinct alleles of the *D1* gene (*d1-3286*, *d1-6039*, *d1-4*, and d1-6016) have been identified [25]. The d1-3286 allele has a 4.5 kb Copia-type retrotransposon inserted at nucleotide position 69 of the *ZmGA3ox2* gene [25]. This insertion disrupts the normal transcriptional and translational processes of the gene, leading to a significant reduction in the production of bioactive gibberellins and ultimately resulting in a dwarf phenotype. The *d1-6039* allele, on the other hand, has a single-nucleotide deletion at position 399 within the first exon [25]. This deletion causes a frameshift mutation, which leads to the premature termination of translation at the 163rd amino acid, severely impairing the function of the encoded enzyme and causing dwarfism. The *d1-6016* allele features a large deletion spanning 2301 bp, encompassing 508 bp of upstream regulatory sequences and 1793 bp of the *ZmGA3ox2* coding region [25]. Such a large-scale deletion disrupts both the regulatory and coding functions of the gene, resulting in a profound dwarf phenotype. The *d1-4* allele contains a 487 bp deletion affecting 389 bp of the first exon and 98 bp of the first intron [25]. This deletion also affects the proper splicing and expression of the gene, leading to a reduction in the activity of the gibberellin 3-beta-dioxygenase enzyme and dwarf plant growth.

In contrast to these previously reported alleles, the 16N125 mutant exhibits a unique mutation pattern. It has a substitution of alanine for proline at position 61 and asparagine for isoleucine at position 123 in the amino acid sequence of the *D1* gene. Through genotypic analysis of the F2 generation and an in-depth investigation of 50 Chinese inbred lines, we determined that the mutation at position 61 is likely the key factor regulating plant height. This is a novel finding, as previous studies on the *D1* gene have not reported such a mutation at this specific position.

This newly identified mutation in 16N125 adds a new dimension to our understanding of the genetic diversity of the *D1* gene. By incorporating this mutation into the *d1-4* allele, we propose an additional locus that contributes to the dwarfing effect of the *d1-4* mutation. This discovery has far-reaching implications for understanding the intricate mechanisms of plant height regulation in maize. It provides a new perspective on how mutations in the *D1* gene can disrupt the gibberellin biosynthesis pathway and ultimately affect plant growth and development.

From a practical perspective, this finding offers potential genetic resources for maize breeding programs. Dwarf or semi-dwarf maize varieties with improved lodging resistance are highly desirable in modern agriculture. Lodging is a major problem in maize production, especially in regions prone to strong winds and heavy rains. By developing varieties with optimized plant heights through the manipulation of the *D1* gene, breeders can enhance the lodging resistance of maize crops, ensuring more stable yields. For example, in regions with high-wind climates, semi-dwarf maize varieties can better withstand strong gusts, reducing the risk of plant lodging and subsequent yield losses [37].

Maize has a long history of cultivation in China since its introduction in the 16th century [38]. Over time, a rich and diverse collection of germplasm resources has been established. However, prior to our study, the status of dwarf genes within common Chinese germplasm resources was not well characterized. By systematically investigating *D1* mutations in 50 widely utilized Chinese inbred lines through Sanger sequencing, we have filled this knowledge gap. Our results indicate that the 61st amino acid of the maize *D1* gene may play a crucial role in influencing plant height under specific conditions.

This information is invaluable for maize breeders in China. It allows them to better understand the genetic background of local inbred lines, which is essential for rational parent selection in breeding programs. For instance, breeders can now identify inbred lines with favorable *D1* gene mutations and use them as parents to develop new varieties with improved plant architecture. By selecting parents with the appropriate *D1* gene mutations, breeders can potentially develop maize varieties that are not only more resistant to lodging but also have enhanced yield potential. In addition, understanding the genetic diversity of the *D1* gene in Chinese inbred lines can help breeders develop varieties that are better adapted to local environmental conditions, such as different soil types and climate patterns [39].

Furthermore, our study also has implications for future research on maize genetics. The identification of this novel mutation in the *D1* gene provides a starting point for further functional studies. For example, researchers can investigate the physiological and biochemical effects of this mutation by analyzing hormone levels, gene expression patterns, and plant growth responses in different genetic backgrounds. Additionally, studying the interaction of the *D1* gene with other genes in the gibberellin biosynthesis pathway and related pathways can provide a more comprehensive understanding of the genetic network governing maize plant height. This knowledge can be applied to develop molecular markers for marker-assisted selection, enabling breeders to efficiently screen for desirable plant height traits in breeding programs.

In conclusion, our study not only validates the effectiveness of BSA-seq in gene mapping but also makes significant contributions to the understanding of the genetic mechanism of maize plant height. The identification of the novel mutation in the *D1* gene and its association with plant height in Chinese inbred lines provide a solid theoretical basis for future maize breeding. This research offers practical guidance for developing maize varieties with improved traits, which is crucial for meeting the growing global demand for food and ensuring sustainable agriculture in the face of climate change. By continuously exploring the genetic diversity of maize and identifying genes associated with important traits, we can develop more resilient and productive maize varieties and contribute to global food security.

4. Materials and Methods

4.1. Plant Materials, DNA Extraction, and Sanger Sequencing

To comprehensively characterize the phenotypic traits of maize, a total of fifty inbred lines and the control group plants were planted in 2020, 2021, and 2022 at the experimental station of Liaoning Academy of Agricultural Sciences located in Shenyang, Liaoning Province, China (42° N, 123° E). Each experimental plot measured 4 m in length and 2.2 m in width, with three rows planted per plot. The row spacing was maintained at 60 cm, while the plant spacing within rows was 30 cm. When it came to the phenotype evaluation process, more than 15 individual plants were carefully selected from the central part of each plot. Specifically, plant height was measured precisely from the ground level all the way up to the top of the plants when they reached the maturity stage. It is worth noting that this measurement method was implemented by referring to the approach proposed by Zou et al. [40], and was further appropriately optimized to ensure its accuracy and suitability for this particular study.

Genomic DNA was extracted following the CTAB protocol. Each sample consisted of 5–6 plants from each accession. The concentration and quality of the extracted DNA were assessed using an ND-1000 spectrophotometer (manufactured by NanoDrop, Wilmington, DE, USA) as well as through electrophoresis on 1.0% agarose gels with lambda DNA as the standard. For the amplification of fragments of *D1*, PCR was carried out in a 25 μL reaction mixture. This mixture contained 2× EXtaq Mix (produced by Takara, Kusatsu, Japan), 0.4 μM of each primer, 0.125 μg of DNA, and an appropriate amount of double-distilled water. The PCR amplification protocol was designed as follows: An initial denaturation cycle was performed at 95 °C for 4 min. Subsequently, 30 cycles were conducted, with each cycle involving denaturation at 95 °C for 30 s, annealing at 57 °C for 30 s, and extension at 72 °C for 2.5 min. The primers utilized for both Sanger sequencing and PCR amplification are presented in Table S2. The PCR products were then sequenced by Biomarker Technologies Corporation (located in Beijing, China).

4.2. SNP Library Construction and High-Throughput Sequencing

The specific length amplified fragment sequencing (SLAF-seq) utilized in our study was carried out with certain adaptations based on the protocol detailed by Sun et al. [41].

The construction of the SLAF library encompassed several vital steps, which were specifically engineered to boost the yield of SLAFs, cut down on repetitive SLAFs, and ensure a balanced distribution of SLAFs, thereby maximizing the sequencing efficiency.

Firstly, a pre-design experiment was implemented to determine the most favorable conditions for constructing the SLAF library. This step was aimed at elevating the overall efficiency of SLAF-seq. Subsequently, in line with the pre-determined plan, genomic DNA was incubated at 37 °C in the presence of *Mse* I (from New England Biolabs, NEB; Ipswich, MA, USA), T4 DNA ligase (NEB), ATP (NEB), and Mse I adapter. After that, the reaction mixture was heat-inactivated at 65 °C and then underwent digestion with the additional restriction enzyme *Alu* I at 37 °C.

Next, polymerase chain reactions (PCRs) were conducted using the diluted restriction–ligation samples, along with dNTPs, Taq DNA polymerase (NEB), and *Mse* I primer containing barcode 1. The PCR products were purified by employing the E.Z.N.A. Cycle Pure Kit (Omega Bio-Tek, Norcross, GA, USA) and then combined in a pool. The pooled samples were then incubated at 37 °C with *Mse* I, T4 DNA ligase, ATP, and a Solexa adapter. Purification was achieved through the use of a Quick Spin column (Qiagen, Hilden, Germany), and the samples were then separated on a 2% agarose gel. Fragments within the size range of 450–500 bp (incorporating indexes and adapters) were isolated by means of a Gel Extraction Kit (Qiagen, Hilden, Germany).

These isolated fragments were then subjected to PCR amplification with Phusion Master Mix (NEB) and Solexa Amplification Primer Mix (Illumina, Inc., San Diego, CA, USA) to integrate barcode 2, following the guidelines provided for Illumina sample preparation. The amplified products were gel-purified, by extracting the DNA fragments sized between 450 and 500 bp, and were then diluted for paired-end sequencing on the Illumina HiSeq 2000 platform (Illumina, Inc., San Diego, CA, USA) at Biomarker Technologies Corporation in Beijing (http://www.biomarker.com.cn (accessed on 7 November 2024)). Subsequently, SNP genotyping and evaluation were carried out.

All SLAF paired-end reads with clear index information were clustered based on sequence similarity, which was detected through one-to-one alignment using BLAT (−tileSize = 10; − stepSize = 5) [42]. Sequences sharing over 90% identity were grouped into one SLAF locus. By making use of the minor allele frequency (MAF) evaluation, alleles were defined within each SLAF.

In the mapping populations of maize, SLAFs having a sequence depth lower than 107 were excluded from the subsequent analysis. SLAFs with 2–4 tags were identified as polymorphic SLAFs and regarded as potential markers. Tags with a total coverage of at least 10× were retained. Sequences were mapped onto the B73 reference genome [43], and SNPs were called using the TASSEL 3.0 GBS pipeline (www.maizegenetics.net/tassel/ (accessed on 9 January 2024)). Missing data were imputed with NPUTE [44].

5. Conclusions

A single-base substitution in the *D1* gene, which encodes gibberellin 3β-dioxygenase, results in recessive dwarfism in the maize mutant 16N125. Unlike the previously reported *d1* alleles, this particular missense mutation is rather subtle. It manages to strike a balance between reducing plant height and minimizing pleiotropic effects. Moreover, as discovered within the Chinese germplasm, it presents a valuable genetic resource for breeding semi-dwarf maize varieties with enhanced lodging resistance.

Supplementary Materials: The following supporting information can be downloaded at: https://www.mdpi.com/article/10.3390/plants14081217/s1, File S1: Amino acid sequence of D1 genes in 50 Chinese inbred lines; Table S1: The sanger sequencing results of the offsprings; Table S2: Primers sequences used in this study.

Author Contributions: P.W. and B.L.: Formal analysis; writing—original draft. X.L.: Conceptualization. Z.L. and H.D.: Methodology, formal analysis, investigation. L.Z.: Methodology, formal analysis, data curation. P.W. and X.L.: Writing—review and editing; supervision; project administration. All authors have read and agreed to the published version of the manuscript.

Funding: This research was supported by the Liaoning Province Special Program for Germplasm Innovation and Store Grain in Technology (2023JH1/10200009) and the China Agricultural Research System (CARS-06-14.5-A3).

Data Availability Statement: The data presented in this study are available upon request from P.W.

Conflicts of Interest: The authors declare no competing interests.

References

1. Brown, L.E.; Chen, S.; Smith, J.D. The role of maize in global food security and nutrition. *Food Policy* **2022**, *108*, 102241.
2. Haley, C. A cornucopia of maize genes. *Nat. Genet.* **2011**, *43*, 87–88. [CrossRef] [PubMed]
3. Xiao, Y.; Jiang, S.; Cheng, Q.; Wang, X.; Yan, J.; Zhang, R.; Qiao, F.; Ma, C.; Luo, J.; Li, W.; et al. The genetic mechanism of heterosis utilization in maize improvement. *Genome Biol.* **2021**, *22*, 148. [CrossRef] [PubMed]
4. Li, W.; Ge, F.; Qiang, Z.; Zhu, L.; Zhang, S.; Chen, L.; Wang, X.; Li, J.; Fu, Y. Maize *ZmRPH1* encodes a microtubule-associated protein that controls plant and ear height. *Plant Biotechnol. J.* **2020**, *18*, 1345–1347. [CrossRef]
5. Le, L.; Guo, W.; Du, D.; Zhang, X.; Wang, W.; Yu, J.; Wang, H.; Qiao, H.; Zhang, C.; Pu, L. A spatiotemporal transcriptomic network dynamically modulates stalk development in maize. *Plant Biotechnol. J.* **2022**, *20*, 2313–2331. [CrossRef]
6. Zheng, Z.P.; Liu, X.H. Genetic analysis of agronomic traits associated with plant architecture by QTL mapping in maize. *Genet. Mol. Res.* **2013**, *12*, 1243–1253. [CrossRef]
7. Ku, L.; Zhang, L.; Tian, Z.; Guo, S.; Su, H.; Ren, Z.; Wang, Z.; Li, G.; Wang, X.; Zhu, Y.; et al. Dissection of the genetic architecture underlying the plant density response by mapping plant height-related traits in maize (*Zea mays* L.). *Mol. Genet. Genom.* **2015**, *290*, 1223–1233. [CrossRef]
8. Batista, R.d.C.M.; Araújo, W.L.; Nunes-Nesi, A.; Siqueira, J.A. Time-to-growth: Photoperiod and photosynthesis make the call. *Trends Plant Sci.* **2024**, *29*, 1159–1161. [CrossRef]
9. Mu, Q.; Guo, T.; Li, X.R.; Yu, J. Phenotypic plasticity in plant height shaped by interaction between genetic loci and diurnal temperature range. *New Phytol.* **2022**, *233*, 1768–1781. [CrossRef]
10. Parihar, A.K.; Hazra, K.K.; Lamichaney, A.; Singh, A.K.; Dixit, G.P. Delineating the role of plant stature towards heat stress tolerance in field pea (*Pisum sativum* L.). *Heliyon* **2023**, *9*, e13876. [CrossRef]
11. Li, Z.Q.; Zhang, H.M.; Wu, X.P.; Sun, Y.; Liu, X.H. Quantitative trait locus analysis for ear height in maize based on a recombinant inbred line population. *Genet. Mol. Res.* **2014**, *13*, 450–456. [CrossRef] [PubMed]
12. Zhou, Z.; Zhang, C.; Zhou, Y.; Hao, Z.; Wang, Z.; Zeng, X.; Di, H.; Li, M.; Zhang, D.; Yong, H.; et al. Genetic dissection of maize plantarchitecture with an ultra-high density bin map based on recombinant inbred lines. *BMC Genom.* **2016**, *17*, 178. [CrossRef] [PubMed]
13. Landis, D.A.; Gardiner, M.M.; van der Werf, W.; Swinton, S.M. Increasing Corn for Biofuel Production Reduces Biocontrol Services in Agricultural Landscapes. *Proc. Natl. Acad. Sci. USA* **2008**, *105*, 20552–20557. [CrossRef] [PubMed]
14. Peng, J.; Richards, D.E.; Hartley, N.M.; Murphy, G.P.; Devos, K.M.; Flintham, J.E.; Beales, J.; Fish, L.J.; Worland, A.J.; Pelica, F.; et al. Green revolution genes encode mutant gibberellin response modulators. *Nature* **1999**, *400*, 256–261. [CrossRef]
15. Khush, G.S. Green revolution: The way forward. *Nat. Rev. Genet.* **2001**, *2*, 815–822. [CrossRef]
16. Sasaki, A.; Ashikari, M.; Ueguchi-Tanaka, M.; Itoh, H.; Nishimura, A.; Swapan, D.; Ishiyama, K.; Saito, T.; Kobayashi, M.; Khush, G.S.; et al. Green revolution: A mutant gibberellin-synthesis gene in rice. *Nature* **2002**, *416*, 701–702. [CrossRef]
17. Lima, M.d.L.A.; de Souza, C.L.; Bento, D.A.V.; de Souza, A.P.; Carlini-Garcia, L.A. Mapping QTL for grain yield and plant traits in a tropical maize population. *Mol. Breed.* **2006**, *17*, 227–239. [CrossRef]
18. Winkler, R.G.; Helentjaris, T. The maize *Dwarf3* gene encodes a cytochrome P450-mediated early step in gibberellin biosynthesis. *Plant Cell* **1995**, *7*, 1307–1317.
19. Multani, D.S.; Briggs, S.P.; Chamberlin, M.A.; Blakeslee, J.J.; Murphy, A.S.; Johal, G.S. Loss of an MDR transporter in compact stalks of maize *br2* and sorghum *dw3* mutants. *Science* **2003**, *302*, 81–84. [CrossRef]
20. Wang, Y.; Li, J. Molecular basis of plant architecture. *Annu. Rev. Plant Biol.* **2008**, *59*, 253–279. [CrossRef]
21. Lawit, S.J.; Wych, H.M.; Xu, D.; Kundu, S.; Tomes, D.T. Maize DELLA proteins dwarf plant8 and dwarf plant9 as modulators of plant development. *Plant Cell Physiol.* **2010**, *51*, 1854–1868. [CrossRef] [PubMed]
22. Phinney, B.O. Growth response of single-gene dwarf mutants in maize to gibberellic acid. *Proc. Natl. Acad. Sci. USA* **1956**, *42*, 185–189. [CrossRef] [PubMed]

23. Spray, C.R.; Kobayashi, M.; Suzuki, Y.; Phinney, B.O.; Gaskin, P.; MacMillan, J. The *dwarf-1 (dt)* Mutant of Zea mays blocks three steps in the gibberellin biosynthetic pathway. *Proc. Natl. Acad. Sci. USA* **1996**, *93*, 10515–10518. [CrossRef] [PubMed]
24. Teng, F.; Zhai, L.; Liu, R.; Bai, W.; Wang, L.; Huo, D.; Tao, Y.; Zheng, Y.; Zhang, Z. *ZmGA3ox2*, a candidate gene for a major QTL, *qPH3.1* for plant height in maize. *Plant J.* **2013**, *73*, 405–416. [CrossRef]
25. Chen, Y.; Hou, M.; Liu, L. The Maize *DWARF1* Encodes a Gibberellin 3—Oxidase and is dual localized to the nucleus and cytosol. *Plant Physiol.* **2014**, *166*, 40–52. [CrossRef]
26. Liu, Y.; Chen, Z.; Zhang, C.; Guo, J.; Liu, Q.; Yin, Y.; Hu, Y.; Xia, H.; Li, B.; Sun, X.; et al. Gene editing of *ZmGA20ox3* improves plant architecture and drought tolerance in maize. *Plant Cell Rep.* **2024**, *43*, 18. [CrossRef]
27. Salas Fernandez, M.G.; Becraft, P.W.; Yin, Y.; Lubberstedt, T. From dwarves to giants? Plant height manipulation for biomass yield. *Trends. Plant Sci.* **2014**, *14*, 454–461. [CrossRef]
28. Andorf, C.M.; Lawrence, C.J.; Harper, L.C.; Schaeffer, M.L.; Campbell, D.A.; Sen, T.Z. The Locus Lookup tool at MaizeGDB: Identification of genomic regions in maize by integrating sequence information with physical and genetic maps. *Bioinformatics* **2010**, *26*, 434–436. [CrossRef]
29. Michelmore, R.W.; Paran, I.; Kesseli, R. Identification of markers linked to disease-resistance genes by bulked segregant analysis: A rapid method to detect markers in specific genomic regions by using segregating populations. *Proc. Nat. Acad. Sci. USA* **1991**, *88*, 9828–9832. [CrossRef]
30. Zhang, K.; Yuan, M.; Xia, H.; He, L.Q.; Ma, J.; Wang, M.X.; Zhao, H.L.; Hou, L.; Zhao, S.Z.; Li, P.C.; et al. BSA-seq and geneticmapping reveals *AhRt2* as a candidate gene responsible for red testa of peanut. *Theor. Appl. Genet.* **2022**, *135*, 1529–1540. [CrossRef]
31. Klein, H.; Xiao, Y.G.; Conklin, P.A.; Govindarajulu, R.; Kelly, J.A.; Scanlon, M.J.; Whipple, C.J.; Bartlett, M. Bulked-segregant analysis coupled to whole genome sequencing (BSA-Seq) for rapid gene cloning in maize. *G3-Genes Genomes Genet.* **2018**, *8*, 3583–3592. [CrossRef] [PubMed]
32. Vogel, G.; Laplant, K.E.; Mazourek, M.; Gore, M.A.; Smart, C.D. A combined BSA-Seq and linkage map approach identifies genomic regions associated with Phytophthora root and crown rot resistance in squash. *Theor. Appl. Genet.* **2021**, *134*, 1015–1031. [CrossRef] [PubMed]
33. Yang, Q.; Zhang, D.F.; Xu, M.L. A sequential quantitative trait locus fine-mapping strategy using recombinant-derived progeny. *J. Integr. Plant Biol.* **2012**, *54*, 228–237. [CrossRef] [PubMed]
34. Zuo, W.L.; Chao, Q.; Zhang, N.; Ye, J.R.; Tan, G.Q.; Li, B.L.; Xing, Y.X.; Zhang, B.Q.; Liu, H.J.; Fengler, K.A.; et al. A maize wall-associated kinase confers quantitative resistance to head smut. *Nat. Genet.* **2015**, *47*, 151–157. [CrossRef]
35. Yu, T.; Zhang, J.; Cao, J. QTL analysis of low temperature tolerance in maize germination by SLAF-seq and BSA technique. *Electron. J. Biotechnol.* **2024**, *70*, 14–22. [CrossRef]
36. Zheng, R.; Deng, M.; Lv, D. Combined BSA-Seq and RNA-Seq reveal genes associated with the visual stay-green of maize (*Zea mays* L.). *Int. J. Mol. Sci.* **2023**, *24*, 17617. [CrossRef]
37. Yang, K.; Tan, J.; Zhang, Q.; Bai, T.; Zhou, S.; Hao, J.; Yu, X.; Zang, Z.; Zhang, D. Research on the genetic improvement effects of lodging resistance-related traits in maize core germplasm. *Agronomy.* **2025**, *15*, 17. [CrossRef]
38. Bray, F. Agriculture. In *Science and Civilisation in China*; Needham, J., Ed.; Cambridge University Press: Cambridge, UK, 1984; Volume 6, Part 2; p. 451.
39. Mathan, J.; Bhattacharya, J.; Ranjan, A. Enhancing crop yield by optimizing plant developmental features. *Development* **2016**, *143*, 3283–3294. [CrossRef]
40. Zou, G.H.; Yan, S.; Zhai, G.W.; Zhang, Z.P.; Zou, J.Q.; Tao, Y.Z. Genetic variability and correlation of stalk yield related traits and sugar concentration of stalk juice in a sweet sorghum (*Sorghum bicolor* L. Moench) population. *Aust. J. Crop Sci.* **2011**, *5*, 1232–1238.
41. Sun, X.W.; Liu, D.Y.; Zhang, X.F.; Li, W.B.; Liu, H.; Hong, W.G.; Jiang, C.B.; Guan, N.; Ma, C.X.; Zeng, H.P.; et al. SLAF-seq: An efficient method of large-scale De novo SNP discovery and genotyping using high-throughput sequencing. *PLoS ONE* **2013**, *8*, e58700. [CrossRef]
42. Kent, W.J. BLAT—The BLAST-like alignment tool. *Genome Res.* **2002**, *12*, 656–664. [PubMed]
43. Zhu, M.; Ma, J.; Liu, X.; Guo, Y.; Qi, X.; Gong, X.; Zhu, Y.; Wang, Y.; Jiang, M. High-resolution map reveals a Ht3-like locus against northern corn leaf blight. *Front. Plant Sci.* **2022**, *13*, 968924.
44. Roberts, A.; McMillan, L.; Wang, W.; Parker, J.; Rusyn, I.; Threadgill, D. Inferring missing genotypes in large SNP panels using fast nearest-neighbor searches over sliding windows. *Bioinformatics* **2007**, *23*, 401–407. [CrossRef] [PubMed]

plants

MDPI

Article

Enhancing Across-Population Genomic Prediction for Maize Hybrids

Guangning Yu [1,2], Furong Li [1,2], Xin Wang [1,2], Yuxiang Zhang [1,2], Kai Zhou [1,2], Wenyan Yang [1,2], Xiusheng Guan [1,2], Xuecai Zhang [3], Chenwu Xu [1,2] and Yang Xu [1,2,*]

1 Key Laboratory of Plant Functional Genomics of the Ministry of Education/Jiangsu Key Laboratory of Crop Genomics and Molecular Breeding, College of Agriculture, Yangzhou University, Yangzhou 225009, China; mx120200746@yzu.edu.cn (G.Y.); lfr953215657yy@163.com (F.L.); seuwangxin@163.com (X.W.); dx120220139@stu.yzu.edu.cn (Y.Z.); mx120200737@yzu.edu.cn (K.Z.); mx120210754@stu.yzu.edu.cn (W.Y.); mx120220781@stu.yzu.edu.cn (X.G.); cwxu@yzu.edu.cn (C.X.)

2 Zhongshan Biological Breeding Laboratory/Jiangsu Co-Innovation Center for Modern Production Technology of Grain Crops, Yangzhou University, Yangzhou 225009, China

3 International Maize and Wheat Improvement Center (CIMMYT), Mexico City 06600, Mexico; xc.zhang@cgiar.org

* Correspondence: xuyang_89@126.com

Abstract: In crop breeding, genomic selection (GS) serves as a powerful tool for predicting unknown phenotypes by using genome-wide markers, aimed at enhancing genetic gain for quantitative traits. However, in practical applications of GS, predictions are not always made within populations or for individuals that are genetically similar to the training population. Therefore, exploring possibilities and effective strategies for across-population prediction becomes an attractive avenue for applying GS technology in breeding practices. In this study, we used an existing maize population of 5820 hybrids as the training population to predict another population of 523 maize hybrids using the GBLUP and BayesB models. We evaluated the impact of optimizing the training population based on the genetic relationship between the training and breeding populations on the accuracy of across-population predictions. The results showed that the prediction accuracy improved to some extent with varying training population sizes. However, the optimal size of the training population differed for various traits. Additionally, we proposed a population structure-based across-population genomic prediction (PSAPGP) strategy, which integrates population structure as a fixed effect in the GS models. Principal component analysis, clustering, and Q-matrix analysis were used to assess the population structure. Notably, when the Q-matrix was used, the across-population prediction exhibited the best performance, with improvements ranging from 8 to 11% for ear weight, ear grain weight and plant height. This is a promising strategy for reducing phenotyping costs and enhancing maize hybrid breeding efficiency.

Keywords: across population; genomic prediction; population structure; maize hybrids

1. Introduction

Maize (*Zea mays*) is one of the most widely grown food crops in the world and plays an important strategic role in ensuring food security and the effective supply of agricultural products [1]. The primary objective of maize breeding is to develop hybrid varieties that exhibit high yield potential across diverse environmental conditions. Taking advantage of heterosis is an effective strategy for enhancing maize yield. Genomic selection (GS) breeding is a technique that utilizes genome-wide markers to predict unknown phenotypes [2], and it is integral to hybrid breeding practices [3,4]. By applying GS technology, breeders are able to estimate the breeding value or phenotypes of all potential crosses of known parents through phenotypic identification and model construction within the training population.

This approach enables them to perform field evaluations on the predicted superior crosses, thereby reducing the corresponding breeding cycle and costs [5].

In the context of GS breeding, predictions are not confined to individuals within a population or to those that are genetically similar to the training population. The across-population prediction presents a cost-effective strategy to leverage genetic information from one population to predict individuals in another, genetically distinct population. While traditional GS relies on training and validating models within genetically similar populations for higher prediction accuracy, across-population predictions enable the utilization of existing datasets, thus minimizing the need for expansive, population-specific data collection. Therefore, investigating effective strategies and prediction accuracy for across-population predictions is essential for enhancing the application of GS technology in breeding practices. Nevertheless, the complexity of across-population prediction is higher than that of within-population prediction, as it involves not only the training population size and marker density but also the kinship between the training and breeding populations. A comprehensive understanding of these influencing factors and their interactions is vital for improving the accuracy of across-population predictions. The genetic relationships between training and breeding populations significantly impact prediction accuracy [6–9].

When the genetic diversity of the training population significantly differs from that of the breeding population, or when the data originate from two entirely distinct populations with low genetic structure correlation, the potential accuracy of the predicted phenotypic values may be limited. The influence of training population composition on prediction accuracy was explored using five double haploid (DH) populations, revealing that combining unrelated populations with opposite linkage phases to the breeding population may result in negative or reduced prediction accuracies [10]. Additionally, the differences in linkage disequilibrium (LD) among different populations, as well as specific allele effects and allele frequencies, could result in low accuracy in across-population predictions [11–14]. Several studies have demonstrated that the accuracy of across-population GS in animals is often low and sometimes even negative [15–17]. Research conducted on wheat populations revealed that genomic prediction across populations demonstrated low accuracy and indicated that it should not be adopted in small populations [18]. Overall, across-population genomic prediction remains a challenge in the field of GS.

To improve the accuracy of across-population genomic prediction, it is essential to examine the optimal size and composition of the training population, particularly when dealing with a large training population [19,20]. The optimal size of the training population is influenced by the heritability of its traits, the genetic relationship between training and breeding populations, and the population structure. Taking population structure and genetic relationships into account can substantially improve the predictive ability of GS models [21,22]. In studies involving rice and wheat populations, the performance of five optimization criteria for selecting training populations was evaluated, indicating that the optimal selection criterion varies depending on the trait architecture and population structure [23]. Using a diversity panel and ten bi-parental crosses within a half-diallel mating design, the across-population predictive abilities for 15 grapevine traits were assessed. The results demonstrated that optimizing the training set with the mean of determination coefficient (CDmean), the mean of prediction error variance (PEVmean), and the mean additive relationship could improve predictive ability for specific traits, thereby highlighting the potential of across-population genomic prediction in grapevine breeding [24].

Large amounts of highly unbalanced historical data are often available in crop breeding programs, particularly when considering data collected across different years and locations. This imbalance is a challenge for accurately training prediction models. During the processing of phenotypic data, researchers often implement strategies such as utilizing Best Linear Unbiased Prediction (BLUP) values to mitigate the phenotypic variations induced by differing environments and temporal factors, thereby reducing the influence of these variables. However, such operations cannot eliminate a series of complex factors such as epistatic effects between genes and the environment, heterosis, and parental inheritance. Therefore, incorporating known parental phenotypes, population structure, and environmental factors into the GS prediction model as covariates or fixed effects can help improve prediction accuracy and provide valuable insights across populations [25–27].

While previous research has conducted across-population predictions in crops such as wheat, rice, and maize, these studies have primarily focused on small populations [10,18,28]. To date, there is a lack of relevant investigations regarding across-population prediction within large maize hybrid populations, particularly when the phenotypic data are derived from BLUP values across various environments and locations. The goal of this study was to optimize across-population genomic prediction for maize hybrids. We proposed a population structure-based across-population genomic prediction (PSAPGP) strategy; this proposed PSAPGP enhances across-population genomic prediction for maize hybrids.

2. Results

2.1. Within- and Across-Population Prediction

The GBLUP and BayesB models were utilized to conduct genomic predictions both within and across populations, with prediction accuracy assessed using Pearson correlation coefficients between predicted and observed phenotypic values. Within the population, prediction accuracies varied between the training and breeding populations across the evaluated traits. Within the training population containing 5.820 maize hybrids (Pop1), the prediction accuracies for three traits, derived from 20 repeated five-fold cross-validations, ranged from 0.775 to 0.940 across various model–trait combinations. In the breeding population consisting of 523 maize hybrids (Pop2), prediction accuracies ranged from 0.502 to 0.712 for all model–trait combinations. Within the training population, the prediction accuracy for the EW and EGW traits showed significant differences between the two models, while PH did not exhibit significant differences. In the breeding population, the prediction accuracy for the three traits did not show significant variation among the different models (Figure 1). For the across-population genomic prediction, the phenotypic values in the breeding population were predicted based on the marker effects estimated from the training population. Figure 2 illustrates the accuracy of these across-population predictions. In the GBLUP model, the prediction accuracies for EW, EGW, and PH were 0.338, 0.313, and 0.431, respectively. In the BayesB model, the prediction accuracies for these traits were 0.321, 0.301, and 0.392, respectively.

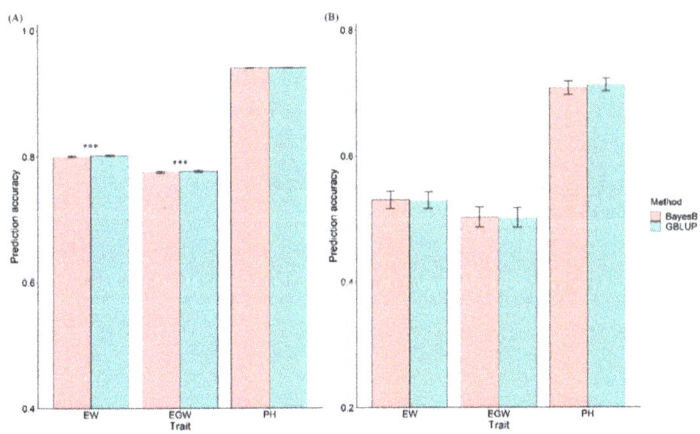

Figure 1. Prediction accuracy within the population under GBLUP and BayesB models. (**A**) Prediction accuracy of the training population; (**B**) prediction accuracy of the breeding population. Note: Asterisks indicate significant differences at $p < 0.001$ (***).

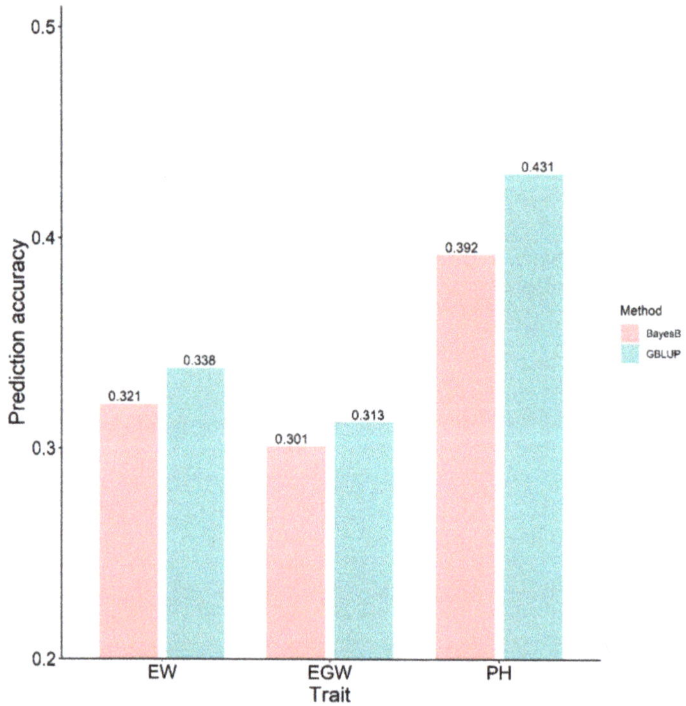

Figure 2. Across-population prediction accuracy using GBLUP and BayesB models.

2.2. Optimization of the Training Population

To enhance genomic prediction accuracy across populations, we optimized the training population based on the genetic relationships between the training and breeding populations. Initially, we constructed a genetic relationship matrix for 6.343 hybrids from Pop1 and Pop2. The genetic distance between each individual in Pop1 and Pop2 increased gradually from left to right (Figure 3A). Consequently, we selected five distinct training population sizes (1000, 2000, 3000, 4000, and 5000) based on the MeanRel and we assessed the prediction accuracy for these varying training population sizes. The results showed that for the traits EW and EGW, both the GBLUP and BayesB models improved the prediction accuracy when the training population size ranged from 1000 to 3000 (Figure 3B,C). Specifically, at a training population size of 2000, EGW exhibited the most obvious increase, with the GBLUP and BayesB models achieving enhancements of 12.0% and 16.9%, respectively. At a training population size of 3000, EW showed the highest increase, with the GBLUP and BayesB models yielding improvements of 6.3% and 11.3%, respectively. In contrast, for PH, the prediction accuracy exhibited the most significant increase when the training population size was between 4000 and 5000, peaking at a training population size of 5000, with the GBLUP and BayesB models achieving enhancements of 2.5% and 8.7%, respectively.

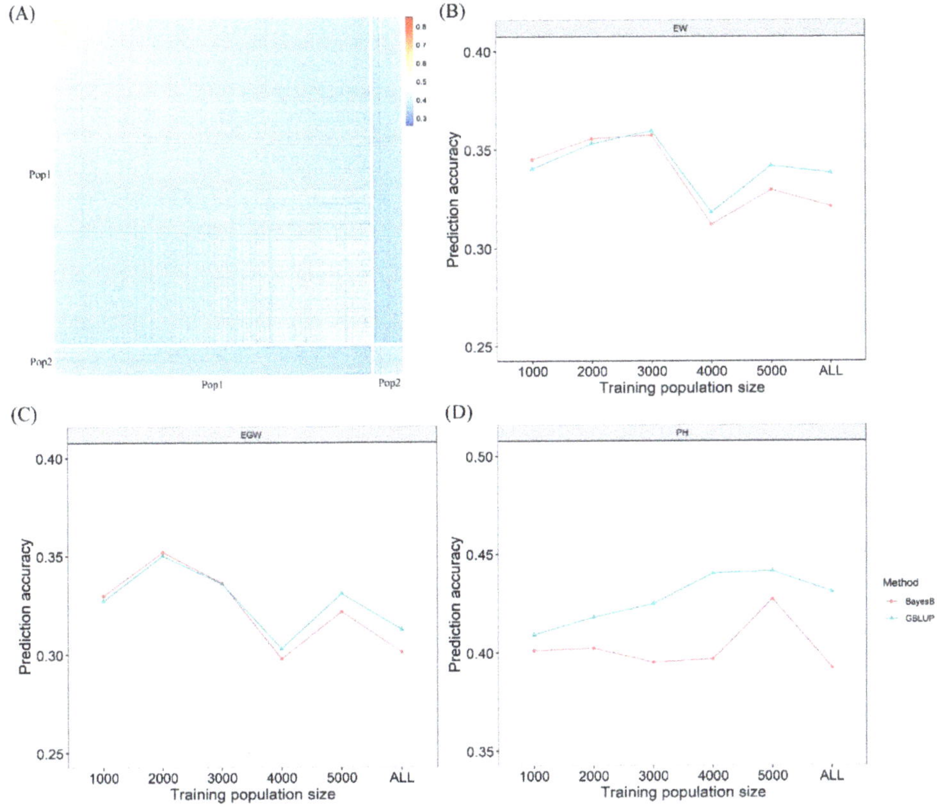

Figure 3. Across-population prediction accuracy of different training population sizes. (**A**) Heat map of kinship matrix for 6.343 maize hybrids; (**B**) prediction accuracy for ear weight; (**C**) prediction accuracy for ear grain weight; (**D**) prediction accuracy for plant height.

2.3. Population Structure-Based Across-Population Genomic Prediction

To investigate the effectiveness of the PSAPGP strategy, we employed PCA, PAM clustering, and Q-matrix analysis to assess the population structure of all hybrids from Pop1 and Pop2. Utilizing genotypic data comprising 58.876 SNPs from 6.343 hybrids, we determined that the optimal number of clusters was six (Figure 4A), after which we conducted the clustering analysis. A three-dimensional PCA plot was generated based on the clustering outcomes, clearly delineating the six categories (Figure 4B). The first six principal components derived from the PCA were subsequently integrated as fixed effects into the GBLUP and BayesB models. For the PAM results, the clustering results were transformed into six columns of data using One-Hot coding, which were also integrated into the GBLUP and BayesB models as fixed effects. The outcomes of the PSAPGP strategy, which integrated PCA and PAM into the GS models, are illustrated in Figure 4C,D. Notably, the integration of PCA and PAM clustering results into the GBLUP model did not yield a significant enhancement in across-population prediction accuracy; in fact, it resulted in a slight decrease. Conversely, when the BayesB model was utilized with the inclusion of PCA results, the prediction accuracy for EW, EGW, and PH exhibited increases of 4.4%, 3.4%, and 2.1%, respectively. In contrast, the integration of PAM results into the BayesB model led to increases in both EW and EGW traits by 3.2%, while PH experienced a minor decline.

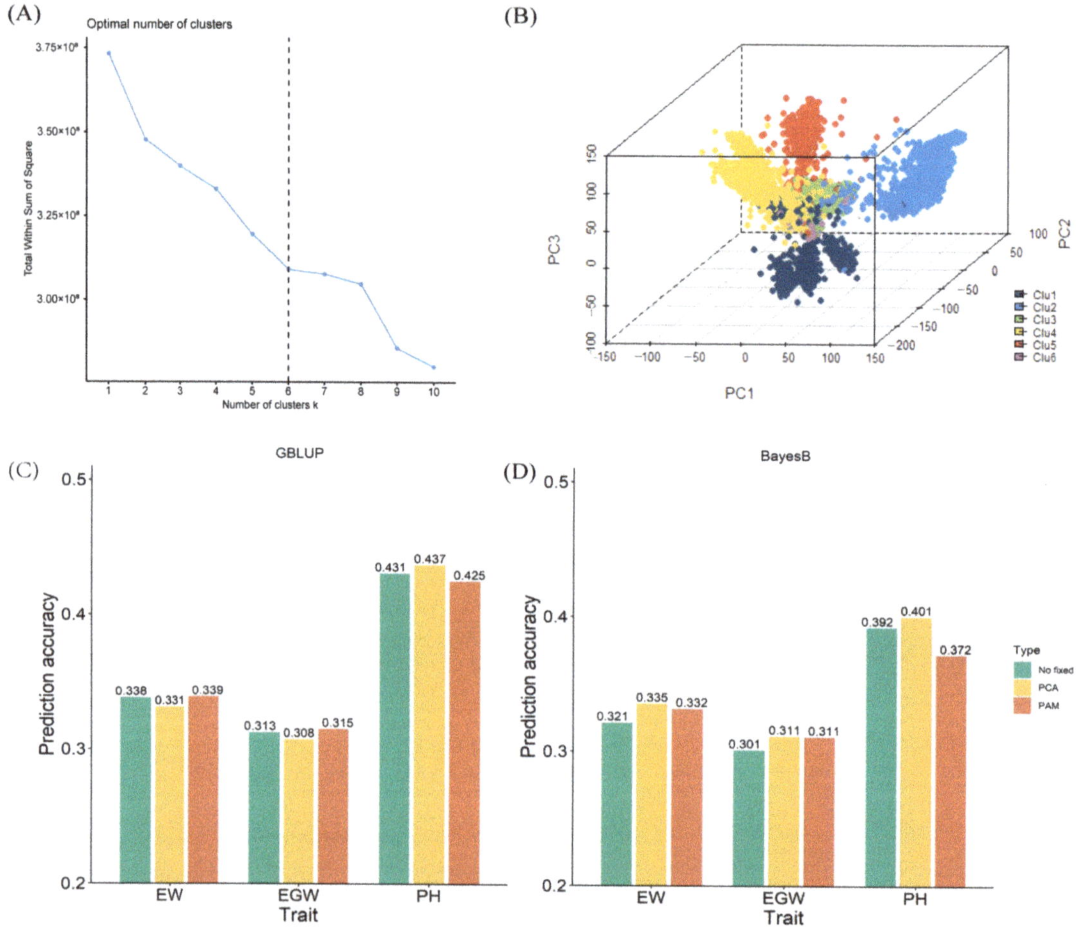

Figure 4. Population structure-based across-population genomic prediction using PCA and PAM as fixed effects; (**A**) scree plot to determine the optimal number of clusters; (**B**) three-dimensional PCA plot of the 6.343 maize hybrids; (**C**) across-population prediction accuracy with the GBLUP model; (**D**) across-population prediction accuracy with the BayesB model.

The population structure Q-matrix of 6.343 hybrids was estimated using ADMIXTURE 1.3.0 software. We evaluated the population structure for K values of 3, 4, 5, 6, and 7 (Figure 5A), and calculated the prediction accuracy of different GS models when different Q-matrices were used as fixed effects. As illustrated in Figure 5, the prediction accuracy of the traits EW, EGW, and PH exhibited significant enhancements across all Q-matrix configurations. Notably, when K = 6 was integrated into the GS model, the prediction accuracies for EW, EGW, and PH improved by 8.4%, 9.2%, and 8.7%, respectively, within the GBLUP model. In the context of the BayesB model, the prediction accuracies for EW, EGW, and PH increased by 9.3%, 10.5%, and 4.9%, respectively. Utilizing the GBLUP method, the maximum observed improvement ratios for EW, EGW, and PH prediction accuracy across various Q-matrix configurations were 8.8%, 9.2%, and 9.3%, respectively. Conversely, under the BayesB method, the maximum improvement ratios for EW, EGW, and PH prediction accuracy across different Q-matrix configurations were 10.3%, 10.5%, and 10.0%, respectively (Figure 5B–D).

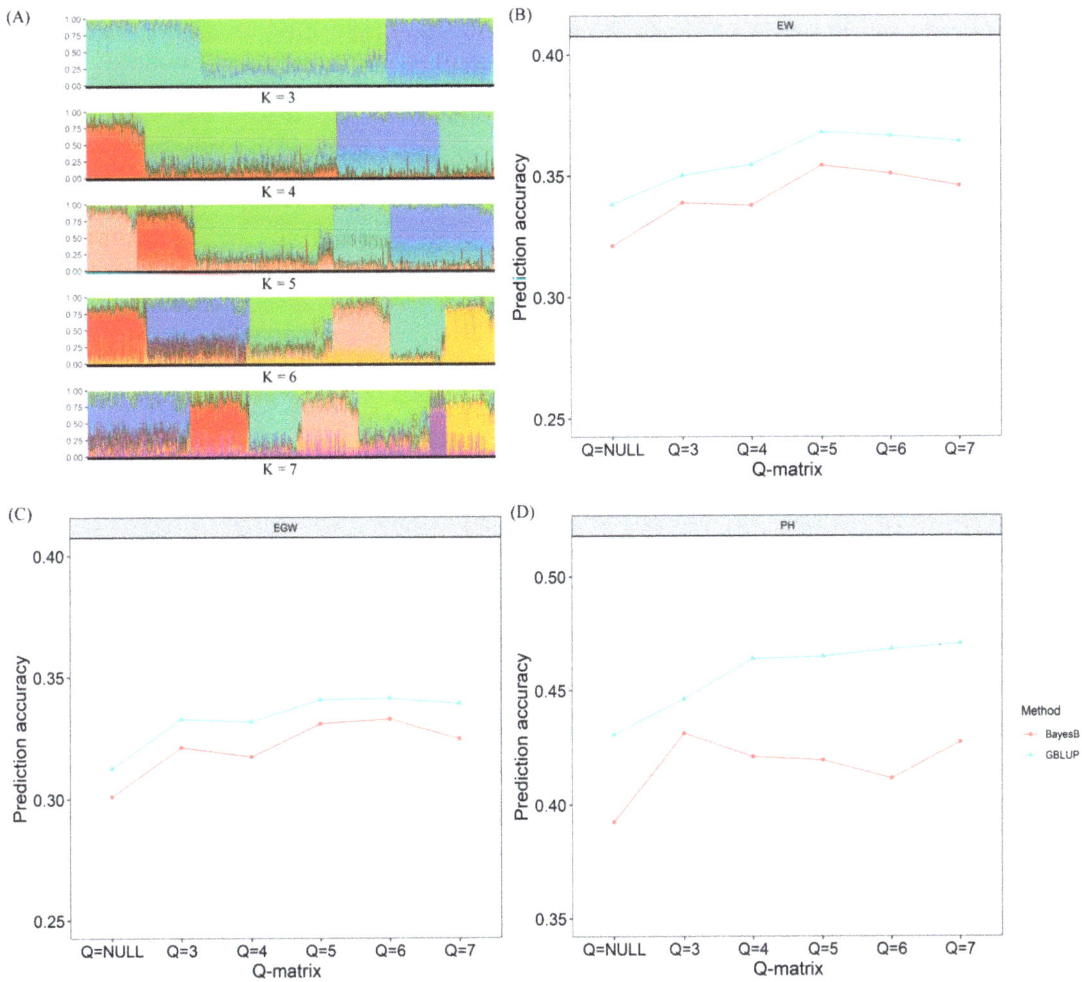

Figure 5. Population structure-based across-population genomic prediction using the Q-matrix as a fixed effect. (**A**) Population structure of the 6.343 maize hybrids at different K values; (**B**) across-population prediction accuracy for ear weight at different K values; (**C**) across-population prediction accuracy for ear grain weight at different K values; (**D**) across-population prediction accuracy for plant height at different K values.

3. Discussion

In this study, we performed genomic predictions both within and across populations using the GBLUP and BayesB models. The lower accuracy of across-population predictions, as compared to within-population predictions, may be attributed to the great genetic distance between the training and breeding populations. When different populations exhibit large genetic distance, the estimation of marker effects can become inconsistent due to variations in alleles, allele frequencies, and linkage phases between the training and breeding populations [29]. Notably, the across-population prediction accuracy of GBLUP was found to be superior to that of BayesB without any optimization of the training population. However, when the training sizes were set at 1000 and 2000, the prediction accuracy of the BayesB method outperformed that of the GBLUP method for the EW and EGW traits. This finding suggests that the selection of the most effective method is contingent upon the genetic relationship between the training and breeding populations.

As breeders integrate GS into ongoing breeding programs, it is essential to carefully select the materials used to construct training populations and optimize the predictive abilities of models derived from these populations. To enhance the accuracy of across-population genomic prediction, previous efforts have focused on optimizing the training population. One study examined the effects of four different population selection methods in a winter wheat panel, finding that selection methods based on PEV outperformed both clustering and random methods [30]. Although employing the PEVmean algorithm to optimize the training population can be beneficial in certain contexts, particularly with small populations, the application of such optimization methods may impose considerable computational demands when larger populations are involved [31]. A method to optimize training sets using kinship, termed MeanRel, has been proposed. This method has demonstrated the potential in enhancing prediction accuracy in the across-population prediction of grapevine [24]. In the present study, we also employed this approach to optimize training populations based on kinship relationships. Our findings revealed varying degrees of improvement across different traits; however, we faced challenges due to the inconsistency in the number of optimal training populations selected for each trait. To address this issue and standardize the selection process, we proposed using the mean value of MeanRel as the cut-off criterion. This approach enabled us to identify the top 2.763 individuals from training populations that were most closely aligned with the breeding population. Under the GBLUP model, the prediction accuracy for the EW, EGW, and PH improved by 7.3%, 9.7%, and 0.07%, respectively. Similarly, under the BayesB model, the prediction accuracy for the same traits increased by 14.4%, 15.5%, and 1.9%, respectively. Notably, both EW and EGW are low-heritability traits. The results suggest that the strategy of optimizing training populations based on kinship has a more obvious advantageous for traits with low heritability.

Research has indicated that incorporating significant SNPs identified through Genome-Wide Association Study (GWAS) as fixed effects can improve the predictive ability of genomic prediction models [32,33]. Utilizing unbalanced phenotypic data collected across varying years or locations, along with markers associated with traits of interest as fixed effects, has been shown to significantly improve the prediction accuracy of key agronomic traits in wheat [30]. In the context of across-population genomic prediction, the accuracy can also be enhanced by integrating significant QTL into the model as fixed effects. The efficacy of across-population predictions is notably improved when the same QTL are present in two populations or when multiple genes are involved [18]. Nevertheless, the strategy of integrating population structure into across-population prediction models has not been extensively investigated. In this study, we propose the PSAPGP strategy to integrate the population structure into prediction models. To study the impact of different population structure evaluation methods on this strategy, we employed PCA, PAM clustering, and Q-matrix analysis to assess the population structure, using two GS models to demonstrate the proposed strategy. Under the GBLUP model, PCA and clustering did not yield significant effects. Conversely, the BayesB model exhibited a modest enhancement in performance. When the Q-matrix was integrated into the model as a fixed effect, both models demonstrated a substantial increase in the accuracy of across-population genomic prediction, with enhancements ranging from 8 to 11%. This improvement may be attributed to the fact that PCA and clustering analyses provide only a basic understanding of population genetic structure, whereas admixture population genetic structure analysis offers additional insights by predicting the optimal number of populations and exploring gene flow between them. Furthermore, it can assess the degree of admixture in individual samples, potentially providing more comprehensive information than the aforementioned methods, thereby positively influencing across-population predictions.

In the breeding of maize hybrids, the vast number of potential crosses makes extensive testing impractical. Identifying superior hybrids from this extensive pool using GS technology presents a long-term challenge. Utilizing public datasets of maize hybrids as a training population for across-population genomic prediction offers a highly cost-effective approach. If these accumulated data can be effectively applied, they will greatly improve the implementation and progress of maize GS breeding. Additionally, this approach of across-population genomic prediction leverages valuable genetic resources from diverse populations to enhance the genetic diversity of hybrids and expand the genetic foundation [28]. We propose a PSAPGP strategy that integrates population structure as a fixed effect in GS models to enhance the accuracy of genomic predictions across populations. Utilizing this strategy, we can further explore the implementation of non-additive effects, such as dominance and epistasis, in genomic prediction models to improve prediction accuracy across populations. Additionally, we may consider integrating multiple population datasets from various sources to expand the size of the training group. In the future, the application of machine learning and deep learning methods is anticipated to facilitate the development of an optimal model for across-population predictions, thereby increasing the accuracy of predictions across diverse populations [34,35]. To successfully implement the across-population GS breeding program, it is essential to establish an open-source GS breeding network along with shared molecular breeding platforms. This network should be capable of offering national agricultural research organizations and small- and medium-sized breeding companies access to advanced and comprehensive breeding technologies. These technologies include universal, efficient, and low-cost genotyping platforms such as target sequencing and liquid chip genotyping, standardized methods for collecting and organizing phenotypic data, and well-established modeling and prediction approaches [36–38]. By sharing breeding platforms and information, and integrating GS breeding strategies, we can continuously update the design and optimization models, improve the accuracy of across-population predictions for maize, and ultimately enhance the breeding efficiency of maize varieties.

4. Materials and Methods

4.1. Training Population of Maize Hybrids

This study utilized two distinct populations of maize hybrids. The first population, referred to as Pop1, serves as the training population, including 5.820 F1 hybrids derived from the cross of 194 maternal inbred lines which were a subset of the maize Complete-diallel plus Unbalanced Breeding-derived Inter-Cross (CUBIC) population and 30 diverse founder paternal lines. The 5.820 F1 hybrids were grown in five locations in 2015 for phenotypic data collection. The analysis focused on three traits: ear weight (EW), ear grain weight (EGW), and plant height (PH). To reduce the impact of environmental variability on the phenotypic data, BLUP values were computed, as referenced in previous studies [39].

4.2. Breeding Population of Maize Hybrids

The second population (Pop2) is a breeding population consisting of 523 hybrids derived from 246 inbred lines belonging to five different heterotic groups, based on a sparse diallel cross design [40]. The maize materials were cultivated in Yangzhou, Jiangsu Province, and Tai'an, Shandong Province, during the year 2017 and 2018. The experimental design employed was a randomized block design with two replications. Each experimental plot had two rows with a length of 3 m and a width of 0.5 m, with 13 plants planted in each row. For each hybrid, five uniform plants were selected to assess three traits: EW, EGW, and PH. The BLUP values for all traits across the two environments were computed for each hybrid using a mixed linear model implemented in the lme4 package in R 4.3.3 [41].

4.3. Genotype Data of Training and Breeding Populations

In the training population, genotype data for 5.820 hybrids were inferred from the genotypes of their respective parent inbred lines. Whole-genome resequencing was conducted on all 1.428 maternal lines and 30 paternal testers, resulting in the identification of 14.8 million single nucleotide polymorphisms (SNPs) [42]. Within the breeding population, 246 inbred lines were genotyped using genotyping by sequencing (GBS), yielding a total of 415.088 SNPs [40]. We screened the genotypes of the 224 inbred lines involved in the 5.820 hybrids, identifying 93.189 SNPs that overlapped with those of the breeding population. Missing genotypes in the hybrids due to heterozygosity of the parents were imputed using Beagle V4.0 [43]. SNPs were filtered with minor allele frequency (MAF) less than 0.05. Ultimately, 58.876 SNPs common to both populations were retained for subsequent analysis.

4.4. Training Population Optimization Methods

The kinship matrix was calculated using the kin function in R/predhy package. Note that the K matrix is partitioned into 2×2 blocks as below:

$$\begin{bmatrix} K_{G11} & K_{G12} \\ K_{G21} & K_{G22} \end{bmatrix} \quad (1)$$

where K_{G11} and K_{G22} represent the kinship matrices for the training population and breeding population, respectively. The off-diagonal elements K_{G12} and K_{G21} represent the kinship between hybrids in the breeding population and those in the training population. We calculated the mean relationship criterion (MeanRel) as described by Brault et al. [24], which represents the average additive relationship between each genotype within the training population and all genotypes in the breeding population.

4.5. The PSAPGP Strategy

The PSAPGP strategy integrates population structure as a fixed effect in across-prediction models. Principal component analysis (PCA), clustering, and Q-matrix analysis were used to assess the population structure of 6.343 maize hybrids from training and breeding populations. The PCA was performed using the R/prcomp function, which enabled the identification of stratified clustering patterns and provided a summary of the genetic variation present across all maize hybrid genotypes. For clustering, we applied the Partitioning Around Medoids (PAM) method, implemented through the R/cluster package. PAM effectively categorizes objects into distinct clusters by minimizing the total dissimilarity between the objects assigned to a cluster and a central representative object, known as the medoid. This method is particularly beneficial for designing training sets, as it not only partitions all objects into clusters but also identifies representative objects for each cluster. The population structure Q-matrix was estimated using ADMIXTURE software, which employs a maximum-likelihood approach to estimate individual ancestries from multi-locus SNP genotype datasets.

The GBLUP and BayesB models were used to demonstrate the effectiveness of PSAPGP. The GBLUP model for PSAPGP is described as:

$$y = Q\beta + Z\gamma_G + \varepsilon \quad (2)$$

where y is the observed vector of the hybrid phenotypic values; Q is an $n \times l$ design matrix for the fixed effect β; Z is an $n \times g$ design matrix for the additive effect γ_G; and ε is a vector of residual errors with an assumed $N(0, I_n\sigma^2)$ distribution. Assume that $\gamma_G \sim N(0, I_g\phi_G^2/g)$, where ϕ_G^2 is the polygenic variance of additive effect. The parameters $\{\beta, \phi_G^2, \sigma^2\}$ were estimated using the restricted maximum likelihood method described in our previous study. Then, the parameters estimated from the training population were used to predict the phenotypic values of the breeding population.

The BayesB model for PSAPGP is described as:

$$y = 1n\mu + Q\beta + Z\gamma_G + \varepsilon \quad (3)$$

where μ is an intercept; Q is a design matrix for the effects of population structure; and β is the corresponding vector of effects, which were treated as 'fixed'. Z is the genotype matrix and γ_G the corresponding vector of marker effects which were treated as random. BayesB assumes that the prior distribution of the variances across markers is a two-component mixture of a point of mass at zero and a scaled-t slab [2]. BayesB were performed using R/BGLR package with all parameters set to default values [44].

5. Conclusions

In this study, we utilized a public dataset of maize hybrid as a training population to facilitate across-population genomic predictions for breeding populations. The optimization of the training population based on kinship yielded varying levels of improvement across different traits, with a more obvious advantage for traits with low heritability. We proposed a PSAPGP strategy, which integrates population structure as a fixed effect in the GS models. PCA, clustering, and Q-matrix analysis were used to assess the population structure. Notably, when the Q-matrix was used, the across-population prediction exhibited the best performance. This marks the first application of a public maize dataset for conducting across-population genomic predictions in maize hybrids. Improving the accuracy of across-population genomic prediction based on training population optimization and the PSAPGP strategy will play a key role in future maize hybrid breeding.

Author Contributions: C.X. and Y.X. designed the research. X.W. and F.L. performed the research. X.G. and W.Y. analyzed the data. Y.Z. and K.Z. developed the software. Y.X. and G.Y. wrote the paper. X.Z., Y.X. and C.X. polished the manuscript. All authors have read and agreed to the published version of the manuscript.

Funding: This work was supported by grants from the Key Research and Development Program of Jiangsu Province (BE2022343), the National Key Research and Development Program of China (2023YFD1202200), the National Natural Science Foundation of China (32170636, 32061143030), Jiangsu Province Agricultural Science and Technology Independent Innovation (CX(21)1003, CX(22)2004), Qing Lan Project of Jiangsu Province, Yangzhou University High-end Talent Support Program, and the Priority Academic Program Development of Jiangsu Higher Education Institutions (PAPD).

Data Availability Statement: The original contributions presented in the study are included in the article, further inquiries can be directed to the corresponding author.

Conflicts of Interest: The authors declare no conflicts of interests.

References

1. Hickey, L.T.; Hafeez, A.N.; Robinson, H.; Jackson, S.A.; Leal-Bertioli, S.C.M.; Tester, M.; Gao, C.X.; Godwin, I.D.; Hayes, B.J.; Wulff, B.B.H. Breeding crops to feed 10 billion. *Nat. Biotechnol.* **2019**, *37*, 744–754. [CrossRef] [PubMed]
2. Meuwissen, T.H.E.; Hayes, B.J.; Goddard, M.E. Prediction of total genetic value using genome-wide dense marker maps. *Genetics* **2001**, *157*, 1819–1829. [CrossRef] [PubMed]
3. Alemu, A.; Åstrand, J.; Montesinos-López, O.A.; Sánchez, J.I.Y.; Fernández-González, J.; Tadesse, W.; Vetukuri, R.R.; Carlsson, A.S.; Ceplitis, A.; Crossa, J.; et al. Genomic selection in plant breeding: Key factors shaping two decades of progress. *Mol. Plant* **2024**, *17*, 552–578. [CrossRef] [PubMed]
4. Xu, Y.; Ma, K.; Zhao, Y.; Wang, X.; Zhou, K.; Yu, G.; Li, C.; Li, P.; Yang, Z.; Xu, C.; et al. Genomic selection: A breakthrough technology in rice breeding. *Crop J.* **2021**, *9*, 669–677. [CrossRef]
5. Nyaga, C.; Gowda, M.; Beyene, Y.; Murithi, W.T.; Burgueno, J.; Toledo, F.; Makumbi, D.; Olsen, M.S.; Das, B.; Suresh, L.M.; et al. Hybrid Breeding for MLN Resistance: Heterosis, Combining Ability, and Hybrid Prediction. *Plants* **2020**, *9*, 468. [CrossRef]
6. Windhausen, V.S.; Atlin, G.N.; Hickey, J.M.; Crossa, J.; Jannink, J.L.; Sorrells, M.E.; Raman, B.; Cairns, J.E.; Tarekegne, A.; Semagn, K.; et al. Effectiveness of Genomic Prediction of Maize Hybrid Performance in Different Breeding Populations and Environments. *G3-Genes Genom Genet.* **2012**, *2*, 1427–1436. [CrossRef]
7. Lund, M.S.; Su, G.; Janss, L.; Guldbrandtsen, B.; Brondurn, R.F. Invited review: Genomic evaluation of cattle in a multi-breed context. *Livest. Sci.* **2014**, *166*, 101–110. [CrossRef]
8. Cericola, F.; Jahoor, A.; Orabi, J.; Andersen, J.R.; Janss, L.L.; Jensen, J. Optimizing Training Population Size and Genotyping Strategy for Genomic Prediction Using Association Study Results and Pedigree Information. A Case of Study in Advanced Wheat Breeding Lines. *PLoS ONE* **2017**, *12*, e0169606. [CrossRef]
9. Badji, A.; Machida, L.; Kwemoi, D.B.; Kumi, F.; Okii, D.; Mwila, N.; Agbahoungba, S.; Ibanda, A.; Bararyenya, A.; Nghituwamhata, S.N.; et al. Factors Influencing Genomic Prediction Accuracies of Tropical Maize Resistance to Fall Armyworm and Weevils. *Plants* **2021**, *10*, 29. [CrossRef]
10. Riedelsheimer, C.; Endelman, J.B.; Stange, M.; Sorrells, M.E.; Jannink, J.L.; Melchinger, A.E. Genomic Predictability of Interconnected Biparental Maize Populations. *Genetics* **2013**, *194*, 493–503. [CrossRef]

11. Zhong, S.Q.; Dekkers, J.C.M.; Fernando, R.L.; Jannink, J.L. Factors Affecting Accuracy from Genomic Selection in Populations Derived From Multiple Inbred Lines: A Barley Case Study. *Genetics* **2009**, *182*, 355–364. [CrossRef] [PubMed]
12. Wientjes, Y.C.J.; Veerkamp, R.F.; Bijma, P.; Bovenhuis, H.; Schrooten, C.; Calus, M.P.L. Empirical and deterministic accuracies of across-population genomic prediction. *Genet. Sel. Evol.* **2015**, *47*, 5. [CrossRef] [PubMed]
13. Wientjes, Y.C.J.; Veerkamp, R.F.; Calus, M.P.L. Using selection index theory to estimate consistency of multi-locus linkage disequilibrium across populations. *BMC Genet.* **2015**, *16*, 87. [CrossRef]
14. Schopp, P.; Müller, D.; Wientjes, Y.C.J.; Melchinger, A.E. Genomic Prediction Within and Across Biparental Families: Means and Variances of Prediction Accuracy and Usefulness of Deterministic Equations. *G3-Genes. Genom. Genet.* **2017**, *7*, 3571–3586. [CrossRef]
15. Hayes, B.J.; Bowman, P.J.; Chamberlain, A.C.; Verbyla, K.; Goddard, M.E. Accuracy of genomic breeding values in multi-breed dairy cattle populations. *Genet. Sel. Evol.* **2009**, *41*, 51. [CrossRef]
16. Lorenz, A.J.; Smith, K.P. Adding Genetically Distant Individuals to Training Populations Reduces Genomic Prediction Accuracy in Barley. *Crop Sci.* **2015**, *55*, 2657–2667. [CrossRef]
17. Pryce, J.E.; Gredler, B.; Bolormaa, S.; Bowman, P.J.; Egger-Danner, C.; Fuerst, C.; Emmerling, R.; Sölkner, J.; Goddard, M.E.; Hayes, B.J. Short communication: Genomic selection using a multi-breed, across-country reference population. *J. Dairy. Sci.* **2011**, *94*, 2625–2630. [CrossRef]
18. Haile, T.A.; Walkowiak, S.; N'Diaye, A.; Clarke, J.M.; Hucl, P.J.; Cuthbert, R.D.; Knox, R.E.; Pozniak, C.J. Genomic prediction of agronomic traits in wheat using different models and cross-validation designs. *Theor. Appl. Genet.* **2021**, *134*, 381–398. [CrossRef]
19. Arruda, M.P.; Brown, P.J.; Lipka, A.E.; Krill, A.M.; Thurber, C.; Kolb, F.L. Genomic Selection for Predicting Head Blight Resistance in a Wheat Breeding Program. *Plant Genome* **2015**, *8*, plantgenome2015.01.0003. [CrossRef]
20. Bentley, A.R.; Scutari, M.; Gosman, N.; Faure, S.; Bedford, F.; Howell, P.; Cockram, J.; Rose, G.A.; Barber, T.; Irigoyen, J.; et al. Applying association mapping and genomic selection to the dissection of key traits in elite European wheat. *Theor. Appl. Genet.* **2014**, *127*, 2619–2633. [CrossRef]
21. Guo, Z.G.; Tucker, D.M.; Basten, C.J.; Gandhi, H.; Ersoz, E.; Guo, B.H.; Xu, Z.Y.; Wang, D.L.; Gay, G. The impact of population structure on genomic prediction in stratified populations. *Theor. Appl. Genet.* **2014**, *127*, 749–762. [CrossRef] [PubMed]
22. Guo, T.T.; Yu, X.Q.; Li, X.R.; Zhang, H.Z.; Zhu, C.S.; Flint-Garcia, S.; McMullen, M.D.; Holland, J.B.; Szalma, S.J.; Wisser, R.J.; et al. Optimal Designs for Genomic Selection in Hybrid Crops. *Mol. Plant* **2019**, *12*, 390–401. [CrossRef] [PubMed]
23. Isidro, J.; Jannink, J.L.; Akdemir, D.; Poland, J.; Heslot, N.; Sorrells, M.E. Training set optimization under population structure in genomic selection. *Theor. Appl. Genet.* **2015**, *128*, 145–158. [CrossRef] [PubMed]
24. Brault, C.; Segura, V.; This, P.; Le Cunff, L.; Flutre, T.; François, P.; Pons, T.; Péros, J.P.; Doligez, A. Across-population genomic prediction in grapevine opens up promising prospects for breeding. *Hortic. Res.* **2022**, *9*, uhac041. [CrossRef]
25. Mulder, H.A. Genomic Selection Improves Response to Selection in Resilience by Exploiting Genotype by Environment Interactions. *Front. Genet.* **2016**, *7*, 178. [CrossRef]
26. Xu, Y.; Zhao, Y.; Wang, X.; Ma, Y.; Li, P.C.; Yang, Z.F.; Zhang, X.C.; Xu, C.W.; Xu, S.Z. Incorporation of parental phenotypic data into multi-omic models improves prediction of yield-related traits in hybrid rice. *Plant Biotechnol. J.* **2021**, *19*, 261–272. [CrossRef]
27. Montesinos-López, O.A.; Herr, A.W.; Crossa, J.; Montesinos-López, A.; Carter, A.H. Enhancing winter wheat prediction with genomics, phenomics and environmental data. *BMC Genom.* **2024**, *25*, 544. [CrossRef]
28. Cui, Y.; Li, R.; Li, G.; Zhang, F.; Zhu, T.; Zhang, Q.; Ali, J.; Li, Z.; Xu, S. Hybrid breeding of rice via genomic selection. *Plant Biotechnol. J.* **2020**, *18*, 57–67. [CrossRef]
29. Bassi, F.M.; Bentley, A.R.; Charmet, G.; Ortiz, R.; Crossa, J. Breeding schemes for the implementation of genomic selection in wheat (spp.). *Plant Sci.* **2016**, *242*, 23–36. [CrossRef]
30. Sarinelli, J.M.; Murphy, J.P.; Tyagi, P.; Holland, J.B.; Johnson, J.W.; Mergoum, M.; Mason, R.E.; Babar, A.; Harrison, S.; Sutton, R.; et al. Training population selection and use of fixed effects to optimize genomic predictions in a historical USA winter wheat panel. *Theor. Appl. Genet.* **2019**, *132*, 1247–1261. [CrossRef]
31. Xu, Y.; Zhang, Y.X.; Cui, Y.R.; Zhou, K.; Yu, G.N.; Yang, W.Y.; Wang, X.; Li, F.R.; Guan, X.S.; Zhang, X.C.; et al. GA-GBLUP: Leveraging the genetic algorithm to improve the predictability of genomic selection. *Brief. Bioinform.* **2024**, *25*, bbae385. [CrossRef] [PubMed]
32. Bernardo, R. Genomewide Selection when Major Genes Are Known. *Crop Sci.* **2014**, *54*, 68–75. [CrossRef]
33. Dang, D.D.; Guan, Y.; Zheng, H.J.; Zhang, X.C.; Zhang, A.; Wang, H.; Ruan, Y.Y.; Qin, L. Genome-Wide Association Study and Genomic Prediction on Plant Architecture Traits in Sweet Corn and Waxy Corn. *Plants* **2023**, *12*, 303. [CrossRef] [PubMed]
34. Gao, P.F.; Zhao, H.N.; Luo, Z.; Lin, Y.F.; Feng, W.J.; Li, Y.L.; Kong, F.J.; Li, X.; Fang, C.; Wang, X.T. SoyDNGP: A web-accessible deep learning framework for genomic prediction in soybean breeding. *Brief. Bioinform.* **2023**, *24*, bbad349. [CrossRef]
35. Montesinos-López, O.A.; Montesinos-López, A.; Pérez-Rodríguez, P.; Barrón-López, J.A.; Martini, J.W.R.; Fajardo-Flores, S.B.; Gaytan-Lugo, L.S.; Santana-Mancilla, P.C.; Crossa, J. A review of deep learning applications for genomic selection. *BMC Genom.* **2021**, *22*, 19. [CrossRef]
36. Xu, Y.B.; Liu, X.G.; Fu, J.J.; Wang, H.W.; Wang, J.K.; Huang, C.L.; Prasanna, B.M.; Olsen, M.S.; Wang, G.Y.; Zhang, A.M. Enhancing Genetic Gain through Genomic Selection: From Livestock to Plants. *Plant Commun.* **2020**, *1*, 100005. [CrossRef]
37. Xu, Y.B.; Zhang, X.P.; Li, H.H.; Zheng, H.J.; Zhang, J.A.; Olsen, M.S.; Varshney, R.K.; Prasanna, B.M.; Qian, Q. Smart breeding driven by big data, artificial intelligence, and integrated genomic-enviromic prediction. *Mol. Plant* **2022**, *15*, 1664–1695. [CrossRef]

38. Li, H.H.; Li, X.; Zhang, P.; Feng, Y.W.; Mi, J.R.; Gao, S.; Sheng, L.L.; Ali, M.; Yang, Z.K.; Li, L.; et al. Smart Breeding Platform: A web-based tool for high-throughput population genetics, phenomics, and genomic selection. *Mol. Plant* **2024**, *17*, 677–681. [CrossRef]
39. Yang, W.; Guo, T.; Luo, J.; Zhang, R.; Zhao, J.; Warburton, M.L.; Xiao, Y.; Yan, J. Target-oriented prioritization: Targeted selection strategy by integrating organismal and molecular traits through predictive analytics in breeding. *Genome Biol.* **2022**, *23*, 80. [CrossRef]
40. Wang, X.; Zhang, Z.L.; Xu, Y.; Li, P.C.; Zhang, X.C.; Xu, C.W. Using genomic data to improve the estimation of general combining ability based on sparse partial diallel cross designs in maize. *Crop J.* **2020**, *8*, 819–829. [CrossRef]
41. Bates, D.; Machler, M.; Bolker, B.M.; Walker, S.C. Fitting Linear Mixed-Effects Models Using lme4. *J. Stat. Softw.* **2015**, *67*, 1–48. [CrossRef]
42. Liu, H.J.; Wang, X.Q.; Xiao, Y.J.; Luo, J.Y.; Qiao, F.; Yang, W.Y.; Zhang, R.Y.; Meng, Y.J.; Sun, J.M.; Yan, S.J.; et al. CUBIC: An atlas of genetic architecture promises directed maize improvement. *Genome Biol.* **2020**, *21*, 20. [CrossRef] [PubMed]
43. Browning, S.R.; Browning, B.L. Rapid and accurate haplotype phasing and missing-data inference for whole-genome association studies by use of localized haplotype clustering. *Am. J. Hum. Genet.* **2007**, *81*, 1084–1097. [CrossRef] [PubMed]
44. Perez, P.; de los Campos, G. Genome-wide regression and prediction with the BGLR statistical package. *Genetics* **2014**, *198*, 483–495. [CrossRef] [PubMed]

plants

MDPI

Article

Semi-Arid Environmental Conditions and Agronomic Traits Impact on the Grain Quality of Diverse Maize Genotypes

Nicolás Francisco Bongianino [1,2,*], María Eugenia Steffolani [1,3], Claudio David Morales [1], Carlos Alberto Biasutti [2] and Alberto Edel León [1,3]

1 Córdoba Food Science and Technology Institute (ICYTAC), National Scientific and Technical, Research Council (CONICET), National University of Córdoba (UNC), Córdoba 5000, Argentina; eusteffolani@agro.unc.edu.ar (M.E.S.); david.morales@unc.edu.ar (C.D.M.); aeleon@agro.unc.edu.ar (A.E.L.)

2 Plant Breeding, College of Agricultural Sciences, National University of Córdoba (UNC), Córdoba 5000, Argentina; biasutti@agro.unc.edu.ar

3 Biological Chemistry, College of Agricultural Sciences, National University of Córdoba (UNC), Córdoba 5000, Argentina

* Correspondence: nicolasbongianino@agro.unc.edu.ar

Abstract: We assessed the impact of environmental conditions and agronomic traits on maize grain quality parameters. The study was conducted using genotypes with distinct genetic constitutions developed specifically for late sowing in semi-arid environments. We evaluated the agronomic, physical, and chemical characteristics of eight maize open-pollinated varieties, six inbred lines, and three commercial hybrids. The yield of the open-pollinated varieties showed a positive correlation with protein content ($r = 0.33$), while it exhibited a negative correlation with the carbohydrate percentage ($r = -0.36$ and -0.42) in conjunction with the inbred lines. The flotation index of the hybrids was influenced primarily by the environmental effect (50.15%), whereas in the inbred lines it was nearly evenly divided between the genotype effect (45.51%) and the environmental effect (43.15%). In the open-pollinated varieties, the genotype effect accounted for 35.09% and the environmental effect for 42.35%. The characteristics of plant structure were associated with grain quality attributes relevant for milling, including hardness and test weight. Inbred lines exhibited significant genotype contributions to grain hardness, protein, and carbohydrate content, distinguishing them from the other two germplasm types. These associations are crucial for specific genotypes and for advancing research and development of cultivars for the food industry.

Keywords: food security; climate change; genetic improvement; variance components; maize quality

1. Introduction

Maize is cultivated in over 160 countries across different agro-climatic zones [1]. This versatile grain can be used to produce food, animal feed, and various industrial applications [2].

Climate change is a widely studied phenomenon caused by increased industrial activity and the emission of greenhouse gases into the atmosphere [3]. This is leading to changes in precipitation and the hydrological cycle [4]. The climate changes could lead to longer droughts and more intense precipitation, affecting groundwater levels due to variations in irrigation demand and water profile fluctuations [5]. The world's food security is at risk due to these changes, which are affecting food production, quality, prices, and supply chains [6].

Reduction in nutrient availability is an aggravating factor in the phenological cycle of crops, which is shortened by all unfavorable conditions (irregular rainfall, higher temperatures, and variations in daylight hours) [6]. As a result, grains from cereals like rice, wheat, and maize have lower yields and a poorer nutritional composition, which has an immediate impact on the quality of food made from them [7–9].

Despite challenging growing conditions, climate adaptation is essential for improving crop performance. Enhancing production stability also requires careful consideration of crop calendar management, strategic variety selection, better irrigation, and waste management [10,11]. Climate change impacts the amount and distribution of rainfall in Argentina's semi-arid central region, influencing maize productivity. Breeding programs must therefore take into account the adaptation of genotypes to these unique conditions [12]. Similarly, evaluating the effects of the environment on genotypes is crucial in these programs, as this can provide information on the assessment of the stability and adaptability of the final genotypes or base populations. In this context, it is important to keep in mind that this may imply that a superior crop variety in one environment may not necessarily perform best in another [13]. Assessing various genotypes for quality traits linked to yield can aid in developing efficient breeding methods for the future [14]. Furthermore, adding value to raw materials will boost the rural economy [15].

There is little information on how climatic conditions and agronomic characteristics affect grain physical and chemical parameters simultaneously evaluated in inbred lines, hybrids, and open-pollinated varieties. Thus, this study aimed to evaluate the effect of environmental conditions and agronomic traits on maize grain quality parameters in inbred lines, hybrids, and open-pollinated varieties developed for late sowing in semi-arid environments.

2. Results and Discussion

2.1. Climate Characterisation

The cumulative rainfall during the 2018 and 2020 crop cycles was below the historical average (489 mm) for this region, with 392 mm and 462 mm. Furthermore, when considered globally, there was a significant decrease in rainfall during the critical periods (cp) (late January–early April) for the genotypes, with 64.25 mm and 71.36 mm. Huang et al. [16] mentioned that understanding crop-specific water requirements enabled effective sowing time management, and alterations in rainfall patterns significantly affected crop yield.

Based on the average temperature fluctuations during the cropping seasons, the highest mean and maximum temperatures occurred in the critical periods of 2018 and 2020. Kothiyal and Kaur [17] highlighted that elevated temperatures significantly affect crop physiological processes, including respiration, photosynthesis, and photoassimilate partitioning. Thus, yield reductions in maize may occur with a temperature increase of up to 1 °C–2 °C.

2.2. Agronomical Traits

Among the hybrid group (Table 1), the P2089 material stood out due to its taller average height of 2.22 m, with the main ear positioned 0.86 m above the soil surface. This cultivar also showed statistical similarities with AX882 and P1815 in ear diameter (4.49 cm, 4.50 cm, and 4.19 cm, respectively). The P2089 genotype outperformed P1815 in grain yield, showing values of 8083.48 kg/ha and 5806.34 kg/ha, respectively. These yields are similar to the average yield from this productive region in recent years (7100 kg/ha) [18]. The inbred lines presented plant heights between 1.71 m (C4B) and 2.06 m (CIM06), a spike insertion height between 0.76 m (C4B) and 1.00 m (CIM06), and a stem diameter between 1.87 cm (B4) and 2.86 cm (BCOT). In addition, the BCOT genotype (5208.57 kg/ha) contrasted significantly with C4B (2422.41 kg/ha) in grain yields. The OPV varieties (C8008 and BlancoM) had a higher average plant size and ear diameter in all growing seasons. Conversely, C980 and C990 displayed some of the highest average yields, with 4451.85 kg/ha and 4469.00 kg/ha, respectively, in comparison with genotypes such as C8008. The latter reached an average yield for all growing seasons of 2624.75 kg/ha. These results are consistent with those published by Sibanda et al. [19], who indicated that values ranged from 2970 kg/ha to 5040 kg/ha for OPVs with a medium yield potential.

Table 1. Growing season average values for agronomic characteristics of hybrids, inbred lines, and open-pollinated varieties.

Genotype	Ger.	Plh (m)	He (m)	Sd (cm)	Ed (cm)	Nr	Ngr	El (cm)	TT (°C d)	Yield (kg/ha)
AX882	H	1.94 ± 0.36 a	0.71 ± 0.19 a	2.03 ± 0.39 a	4.50 ± 0.55 a	15.00 ± 1.68 a	31.20 ± 6.22 a	14.89 ± 2.31 a	946.20 ± 81.23 a	7071.10 ± 4741.44 ab
P1815	H	2.02 ± 0.43 a	0.79 ± 0.35 b	2.13 ± 0.46 a	4.19 ± 0.58 a	14.88 ± 1.96 a	29.85 ± 8.74 a	14.01 ± 2.98 a	1044.05 ± 83.75 b	5806.34 ± 3896.20 a
P2089	H	2.22 ± 0.50 b	0.86 ± 0.29 b	2.14 ± 0.49 a	4.49 ± 0.61 a	15.13 ± 2.13 a	32.85 ± 10.76 a	14.85 ± 3.90 a	1047.28 ± 73.53 b	8083.48 ± 5307.75 b
B4	L	1.80 ± 0.23 ab	0.87 ± 0.22 ab	2.44 ± 0.45 a	4.03 ± 0.56 a	14.25 ± 1.61 a	25.45 ± 1.24 ab	14.09 ± 2.11 a	979.28 ± 73.26 d	3297.58 ± 1137.80 ab
BCOT	L	2.00 ± 0.59 ab	1.00 ± 0.25 b	2.85 ± 0.18 c	3.88 ± 0.72 a	13.55 ± 1.65 a	34.00 ± 0.00 c	14.63 ± 3.67 a	854.63 ± 15.79 ab	5208.57 ± 3666.23 b
BL04	L	1.87 ± 0.51 ab	0.96 ± 0.12 ab	2.47 ± 0.15 b	3.72 ± 0.61 a	13.8 ± 1.48 a	25.50 ± 1.84 ab	13.80 ± 3.26 a	890.78 ± 57.53 c	3183.14 ± 1503.38 ab
BulkASC	L	1.97 ± 0.47 ab	0.95 ± 0.13 ab	2.46 ± 0.04 b	3.95 ± 0.85 a	14.25 ± 2.22 a	25.70 ± 1.27 ab	12.62 ± 1.80 a	879.90 ± 32.79 bc	3035.02 ± 1364.56 ab
C4B	L	1.71 ± 0.25 a	0.76 ± 0.17 a	2.52 ± 0.08 b	3.42 ± 0.34 a	12.40 ± 0.57 a	21.10 ± 2.40 a	11.98 ± 2.22 a	957.95 ± 26.33 d	2422.41 ± 738.69 a
CIM06	L	2.06 ± 0.46 b	0.89 ± 0.18 ab	2.59 ± 0.08 bc	3.63 ± 0.78 a	13.65 ± 1.68 a	27.90 ± 2.69 bc	15.41 ± 3.75 a	846.53 ± 25.14 a	4064.77 ± 2843.87 ab
BlancoM	OPV	2.27 ± 0.59 ab	1.17 ± 0.36 c	2.43 ± 0.50 bc	4.34 ± 0.52 b	12.50 ± 0.53 a	27.50 ± 7.34 b	15.58 ± 3.49 a	1268.73 ± 125.56 b	3188.54 ± 2110.93 ab
C6006	OPV	2.09 ± 0.49 ab	0.99 ± 0.27 ab	2.26 ± 0.49 abc	4.01 ± 0.33 a	12.85 ± 1.50 a	25.40 ± 4.13 ab	14.35 ± 2.08 a	1111.28 ± 68.29 a	3940.18 ± 2004.07 ab
C8008	OPV	2.25 ± 0.56 b	1.13 ± 0.36 bc	2.50 ± 0.48 c	4.35 ± 0.50 b	13.48 ± 1.51 a	25.90 ± 8.07 ab	15.08 ± 2.67 a	1300.15 ± 104.91 b	2624.75 ± 1736.43 a
C900	OPV	2.10 ± 0.44 ab	1.01 ± 0.27 abc	2.19 ± 0.48 ab	3.98 ± 0.39 a	12.83 ± 1.53 a	25.95 ± 7.80 ab	14.24 ± 2.88 a	1111.50 ± 70.11 a	3574.68 ± 1897.84 ab
C980	OPV	2.10 ± 0.41 ab	0.96 ± 0.30 ab	2.37 ± 0.47 abc	4.11 ± 0.34 ab	13.10 ± 0.95 a	26.50 ± 2.83 ab	15.10 ± 2.27 a	1121.73 ± 77.85 a	4451.85 ± 2168.74 b
C990	OPV	2.17 ± 0.44 ab	1.03 ± 0.28 abc	2.23 ± 0.39 abc	4.05 ± 0.44 ab	13.25 ± 0.92 a	25.05 ± 8.08 ab	14.19 ± 2.82 a	1137.95 ± 70.54 a	4469.00 ± 2014.61 b
CandelariaINTA	OPV	2.13 ± 0.48 ab	0.95 ± 0.31 a	2.28 ± 0.45 abc	4.19 ± 0.36 ab	13.65 ± 1.09 a	28.55 ± 11.85 ab	14.89 ± 3.36 a	1125.26 ± 62.03 a	4244.14 ± 2731.87 ab
LealesINTA	OPV	2.01 ± 0.37 a	0.96 ± 0.26 ab	2.08 ± 0.42 a	4.25 ± 0.38 ab	14.14 ± 1.02 a	23.60 ± 6.48 a	13.94 ± 2.81 a	1131.66 ± 60.92 a	3077.72 ± 1846.28 ab

Average values (± standard deviation) where different letters on each column and genotype group indicate statistical differences (Tukey's HSD test, $p \leq 0.05$). Ger., type of germplasm; H, hybrid; L, inbred line; OPV, open-pollinated varieties; Plh, plant height; He, height of insertion of the main ear; Sd, stem diameter; Ed, ear diameter; Nr, number of rows; Ngr, number of grains per row; El, ear length; TT, thermal time to flowering.

2.3. Grain Physical–Chemical Characteristics

Regarding the weight of 1000 grains (Table 2) analyzed in the hybrids, the genotype AX882 (310.16 g) showed a significant superiority over P1815 (243.39 g). Furthermore, the genotype P1815 was superior to the other hybrids in terms of ash content (1.74%). The L cultivars B4 and BL04 had the lowest grain hardness, as indicated by their highest flotation index values (62.13% and 64.5%, respectively). Furthermore, the genotype B4 exhibited the highest W1000. However, these same lines diverged significantly in grain chemical composition. Both B4 and BL04 obtained the lowest protein content values (9.42% and 9.66%, respectively), yet they showcased the highest expression in terms of carbohydrate percentage (84.98% and 85.55%). Sharma and Carena [15] pointed out that the chemical composition, including protein, oil and total starch content, is predominantly influenced by additive gene action. Furthermore, they observed a significant correlation between parental and offspring genotypes for these traits, which intensifies under extreme environmental conditions. This suggests that data from inbred lines could predict hybrid performance across varying environments.

The open-pollinated varieties (OPVs) BlancoM and C8008 demonstrated a superior performance in several key metrics. BlancoM exhibited the highest flotation index at 79.5%, followed by C8008 with 60.13%. In terms of grain weight, BlancoM also had the highest value for a weight of 1000 grains at 342.32 g. However, the C8008 variety showed the lowest test weight values, with C8008 at 80.84 kg/HL as detailed in Table 2. Regarding nutritional content, the OPVs displayed protein contents ranging from 9.05% to 10.67%, lipids contents between 4.27% and 5.74%, and carbohydrates between 81.80% and 84.80%. Ash content showed minimal variation across all harvest years, with values from 1.64% to 1.79%. Kljak et al. [20] mentioned that a higher number of floating grains indicates lower hardness, vitreousness percentage, and breakage susceptibility. The hardness of the grain is explained by the arrangement of starch granules within the protein matrix of the grain endosperm. The mealy endosperm is characterized by loosely organized granules in a thinner protein matrix with multiple air-filled spaces, while glassy endosperm contains tightly organized granules. Environmental factors before harvest and post-harvest conditions such as handling, transport, drying, storage, and processing can influence grain hardness; however, it is primarily determined by the genetic expression unique to each variety [21]. Nguma et al. [22] noted that certain types of zein are associated with grain hardness; specifically, jagged grains contain higher levels of γ-zein synthesized in the mealy fraction of the endosperm's inner core [23].

Grain physical aspects, such as the flotation index and weight of 1000 grains (W1000), were negatively associated with mean yield across all growing seasons, as indicated in Table 2. The flotation index was correlated with yield, showing r coefficients of −0.39 for L and −0.28 for OPV. Additionally, the W1000 values ranged between −0.62 and −0.83 for the three material types examined. Nemati et al. [24] reported that the 1000 grain weight (TGW) had no significant effect on yield in early planting when a longer growing season is evident. This contrasts with Jamshidian et al. [25], who suggested that indirect selection for traits like TGW can improve ear yield, especially in early generations of dent maize hybrids. Cerrudo et al. [26] further highlighted that maize grain hardness was independent of yield but rather dependent on the availability of photoassimilates per unit grain. They proposed that a combination of increased photosynthetic activity and reduced grain number due to growth restriction could ultimately reduce yield while increasing grain hardness.

Table 2. Growing season average values for physical–chemical grain characteristics and their correlations with yield in hybrids, inbred lines, and open-pollinated varieties.

Genotype	Ger.	FI (%)	Hardness	W1000 (g)	TW (kg/HL)	Protein (%)	Lipids (%)	Ch (%)	Ash (%)
AX882	H	83 ± 14.89 a	Soft	310.16 ± 48.86 b	82.83 ± 4.81 a	8.05 ± 1.3 a	4.68 ± 0.42 a	85.62 ± 1.43 a	1.65 ± 0.09 a
P1815		86.63 ± 11.53 a	Soft	243.39 ± 60.23 a	83.63 ± 7.68 a	7.71 ± 0.57 a	5.03 ± 1.72 a	85.51 ± 1.7 a	1.74 ± 0.11 b
P2089		90.38 ± 6.86 a	Very Soft	289.66 ± 63.13 ab	82.13 ± 6.54 a	7.38 ± 0.84 a	4.71 ± 0.24 a	86.29 ± 0.98 a	1.62 ± 0.05 a
Yield		n/s		−0.83 **	n/s	n/s	n/s	n/s	−0.68 **
B4	L	62.13 ± 15.98 b	Soft	325.09 ± 34.68 c	86.13 ± 5.29 a	9.42 ± 1.12 a	3.92 ± 0.7 a	84.98 ± 1.28 c	1.68 ± 0.15 ab
BCOT		22.25 ± 9.54 a	Hard	270.84 ± 51.4 b	93.4 ± 0.28 a	12.05 ± 1.59 b	4.54 ± 2.94 a	81.56 ± 1.41 a	1.84 ± 0.09 b
BL04		64.5 ± 23.74 b	Soft	271.44 ± 38.76 b	87.8 ± 0.28 a	9.66 ± 0.33 a	3.28 ± 0.51 a	85.55 ± 0.8 c	1.51 ± 0.04 a
BulkASC		25.75 ± 20.17 a	Hard	262.73 ± 62.26 b	85.25 ± 9.77 a	12.19 ± 1.13 b	3.42 ± 1.8 a	82.63 ± 0.72 ab	1.76 ± 0.05 ab
C4B		21.88 ± 18.88 a	Hard	189.59 ± 55.06 a	95.5 ± 0.14 a	12.53 ± 1.09 b	3.31 ± 1.38 a	82.50 ± 0.56 ab	1.66 ± 0.27 ab
CIM06		19.75 ± 10.9 a	Hard	262.44 ± 82.94 b	85.45 ± 8.17 a	11.15 ± 0.21 ab	3.77 ± 0.09 a	83.45 ± 0.12 b	1.63 ± 0.01 ab
Yield		−0.39 *		−0.62 **	0.63 **	n/s	0.76 **	−0.42 *	n/s
BlancoM	OPV	79.5 ± 16.78 c	Soft	342.32 ± 48.42 c	82.08 ± 6.92 ab	9.18 ± 0.75 a	4.27 ± 1.1 a	84.80 ± 1.6 c	1.76 ± 0.25 a
C6006		30.63 ± 23.35 a	Hard	279.03 ± 37.51 ab	87.35 ± 4.67 c	9.45 ± 1.21 ab	4.44 ± 0.93 a	84.42 ± 1.27 bc	1.69 ± 0.09 a
C8008		60.13 ± 11.31 b	Intermediate	303.41 ± 66.67 b	80.84 ± 7.63 a	9.78 ± 0.81 ab	4.6 ± 1.1 ab	83.94 ± 1.66 bc	1.68 ± 0.09 a
C900		39.5 ± 16.86 a	Intermediate	277.96 ± 46.79 ab	86.8 ± 6.1 bc	10.19 ± 0.88 ab	5.01 ± 0.66 ab	83.04 ± 1.24 ab	1.75 ± 0.08 a
C980		38 ± 14.07 a	Intermediate	254.88 ± 41.76 a	87.35 ± 5.64 c	9.77 ± 0.74 ab	4.44 ± 0.89 a	84.07 ± 0.73 bc	1.71 ± 0.12 a
C990		45.63 ± 21.61 ab	Intermediate	270.66 ± 43.05 ab	86.65 ± 6.9 bc	9.86 ± 0.64 ab	5.11 ± 1.56 ab	83.31 ± 1.88 abc	1.72 ± 0.12 a
CandelariaINTA		40.38 ± 16.05 a	Intermediate	263.59 ± 34.05 a	84.38 ± 5.31 abc	9.05 ± 0.98 a	4.52 ± 0.63 a	84.79 ± 0.82 c	1.64 ± 0.05 a
LealesINTA		39.75 ± 24.71 a	Intermediate	257.59 ± 39.43 a	84.14 ± 5.7 abc	10.67 ± 1.76 b	5.74 ± 1.52 b	81.80 ± 2.55 a	1.79 ± 0.17 a
Yield		−0.28 *		−0.8 **	n/s	0.33 **	n/s	−0.36 **	n/s

Average values (±standard deviation) where different letters on each column and genotype group indicate statistical differences (Tukey's HSD test, $p \leq 0.05$). * and ** indicate significance at 95 and 99%, respectively. n/s means not significant. Ger., type of germplasm; H, hybrid; L, inbred line; OPV, open-pollinated varieties; FI, flotation index; W1000, weight of a thousand grains; TW, test weight; Ch, carbohydrates.

The protein content in OPV showed a positive correlation with yield (r = 0.33). This suggests that increasing nitrogen uptake or maintaining high nitrogen partitioning within the grain can lead to higher protein [27]. However, Šeremešić et al. [28] found that modern hybrids exhibited a different response of grain protein content to plant density and environment compared to older varieties. Inbred lines and OPV exhibited negative correlations between carbohydrate content and yield, with r correlation coefficients of −0.42 and −0.36, respectively (Table 2). Amegbor et al. [29] reported significant positive associations between grain yield and protein content, while starch content was negatively associated, which is in line with current trends. Additionally, lipid content showed a strong positive correlation for L, with an r-value of 0.76. Ramana Reddy et al. [30] also demonstrated a positive relationship between grain yield and lipid content for both inbred lines and hybrid crosses, with values of 0.35 and 0.52, respectively. However, other researchers have observed that the development of increasingly productive hybrid corn varieties has led to a decrease in nutritional value, particularly in protein and lipid content [31].

The thermal time required to reach the phenological flowering stage was found to be significantly and positively correlated with grain physical parameters, such as the weight of 1000 grains (r = 0.37, p = 0.0001) and the flotation index (r = 0.29, p = 0.0016). Conversely, a negative correlation was observed with protein content (r = −0.30, p = 0.0011). Butts-Wilmsmeyer et al. [32] observed that the accumulation of storage materials such as proteins and starch during grain development is influenced by temperature and soil moisture, which in turn affects yield and grain composition. Similarly, Chen et al. [33] demonstrated that drought stress leads to reduced starch accumulation in the endosperm, significantly contributing to a decrease in grain weight [34]. Jahangirlou et al. [35] highlighted that resource availability, particularly water and nitrogen, significantly impacts germ growth and the enzymatic activity involved in lipid biosynthesis. They found that water availability during the critical period (cp) was positively correlated with both the grain test weight (r = 0.21, p = 0.0290) and lipid content (r = 0.29, p = 0.0018). Additionally, plants undergoing flowering have mechanisms to withstand stress by accelerating this phase to expedite flower and seed production [36]. Moreover, certain plants have evolved drought-tolerance mechanisms that enhance water-use efficiency [37].

Maximum temperatures were found to be positively correlated with W1000 (r = 0.62, $p < 0.0001$ and FI (r = 0.36, p = 0.0001), while showing a negative correlation with TW (r = −0.39, $p < 0.0001$). These temperatures also correlated with a higher percentage of carbohydrates (r = 0.49, p = 0.0001) and a lower lipid content (r = −0.54, $p < 0.0001$). Significant correlations were also observed between the grain's physical characteristics and its chemical composition. Notably, protein content was negatively correlated with both the flotation index (r = −0.72, $p < 0.0001$) and W1000 (r = −0.27, p = 0.0033). Lipid content showed an inverse relationship with W1000 (r = −0.36, p = 0.0001) and a direct relationship with TW (r = 0.51, $p < 0.0001$). Carbohydrate content exhibited significant correlations with all three physical parameters: a negative correlation with hectoliter weight (r = −0.39, $p < 0.0001$), and positive correlations with P1000 (r = 0.47, $p < 0.0001$) and IF (r = 0.69, $p < 0.0001$). Narváez-González et al. [38] indicated that the compactness of endosperm cell bodies determines the physical properties of the grain, including weight, size, hard endosperm percentage, protein content, and pericarp thickness.

The associations between the physical–chemical parameters of the grain and the agronomic characteristics of the genotypes suggest that variability can be largely explained by two principal components: PC1 (52.1%) and PC2 (22.2%), accounting for 85.7% of the total variation (Figure 1). The weight of W1000 and the flotation index are inversely related to all of the structural parameters of the plant. Among the chemical parameters, protein content showed the strongest relationship with genotype-specific traits, particularly those related to vegetative structure and ear characteristics. Conversely, carbohydrate percentage showed a negative correlation with these agronomic traits, while lipid content displayed a similar pattern to that of proteins. Grain hardness, an intrinsic property of genotypes, is negatively associated with starch content but positively correlated with amylose levels [39].

However, environmental factors can influence the starch biosynthetic pathway, affecting enzymes involved in substrate production, the elongation of α-1,4-glucan chains, branching, and maintenance of the granules' crystalline structure [39,40].

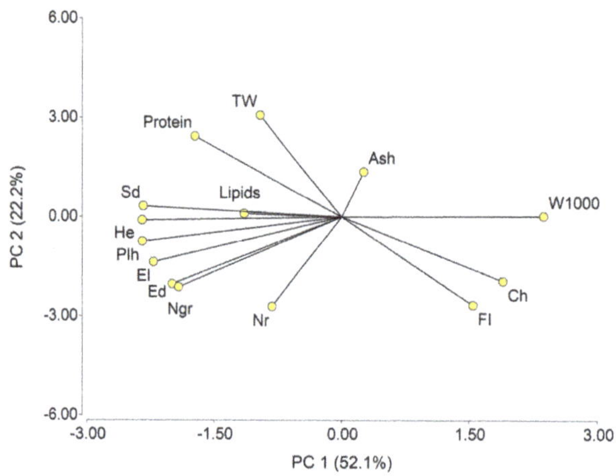

Figure 1. Principal component analysis for grain physical–chemical parameters and agronomic characteristics. W1000, weight of a thousand grains; TW, test weight; FI, flotation index; Ch, carbohydrates; Plh, plant height; He, height of insertion of the main ear; Sd, stem diameter; El, ear length; Ed, ear diameter; Nr, number of rows; Ngr, number of grains per row.

2.4. Variance Components for Grain Physical–Chemical Characteristics

Firstly, it was observed that the flotation index varied among the materials evaluated (Table 3). Environmental effects accounted for 50.15% of the variability in flotation index expression for H, while the variability was nearly evenly split between genotype effects (45.51%) and environmental effects (43.15%) for inbred lines. In OPVs, genotype effects represented 35.09% of the variability and environmental effects accounted for 42.35%. This important incidence of the genotype effect on the indirect expression of grain hardness (flotation index) coincides—although with a lower value—with that reported by Caballero-Rothar et al. [41], who indicated a contribution of 88% of the total variability. For W1000, the explanation of the variability between materials was very similar, since the environmental effect obtained the highest participation, with values between 84.33% and 92.80%. Similarly, the effect of genotype and the GxE interaction was very small or null for TW; thus, the environmental effect obtained a high percentage in the three types of germplasm.

The expression of protein character in the lines was influenced almost equally by genotype effects and genotype–environment interactions, with variance percentages of 45.9% and 41.77%, respectively (Table 3). In contrast, environmental factors played a more dominant role in the variance components for open-pollinated varieties (OPVs), despite having a significant 58.54% influence on hybrids. Previous studies [42,43] have highlighted the substantial impact of environmental factors on protein expression. Regarding lipid content, environmental factors and their interaction with genotype were the most influential, with the genotype explaining only a minor percentage of the variability in OPVs. The significance of the environmental–genotype interaction on lipid content is supported by findings from Fang et al. [44]. While genetics plays a major role in determining lipid concentration in maize grain, environmental conditions and agronomic practices such as fertilization and planting density can also induce variations [45]. Ndlovu et al. [46] noted that nitrogen levels in soil could affect protein content and the saturated/unsaturated fatty acid ratio in grain lipids. Ngaboyisonga and Njoroge [47] found that certain inbred lines produced more stable offspring with higher grain hardness, protein quantity, and quality under low-nitrogen drought conditions. The influence of the male gamete (pollen) on lipid content suggests a paternal effect without impacting seed weight, unlike protein content which is solely affected by maternal factors [48–50]. Ash content was represented by genotype effects (29.95%) and environmental effects (35.45%) in hybrids, while in inbred lines and OPVs these components poorly explained this trait (Table 3). The effect of the environment on the material evidenced less stability in expressing its attributes when a character demonstrated a significant GxE interaction, as noted by Etiro et al. [51].

Table 3. Variance components for physical–chemical grain parameters evaluated in hybrids, inbred lines, and open-pollinated varieties for all growing seasons.

Genotype	VC (%)	FI (%)	W1000 (g)	TW (kg/HL)	Protein (%)	Lipids (%)	Ch (%)	Ash (%)
Hybrid	G	0.94	4.19	0.00	5.42	0.00	0.00	29.95
	E	50.15	84.33	88.40	58.54	24.67	45.66	35.45
	GxE	16.36	6.79	0.00	4.06	69.10	33.35	4.21
	Residual	32.55	4.69	11.60	31.99	6.23	20.98	30.39
Line	G	45.51	3.09	0.00	45.90	0.00	61.89	26.77
	E	43.15	87.59	85.78	7.66	47.68	21.88	1.96
	GxE	6.94	8.85	13.04	41.77	49.50	10.16	7.54
	Residual	4.40	0.47	1.18	4.66	2.82	6.06	63.74
OPV	G	35.09	2.49	8.96	12.06	7.69	22.74	0.00
	E	42.35	92.80	68.83	34.17	47.57	37.45	0.00
	GxE	18.13	3.57	10.91	20.07	23.17	13.67	22.64
	Residual	4.44	1.15	11.30	33.70	21.57	26.13	77.36

VC, variance component; FI, flotation index; W1000, weight of a thousand grains; TW, test weight; Ch, carbohydrates; G, genotypic effect; E, environmental effect; GxE, genotype–environment interaction effect.

Özdemir and Sade [52] stated that trait variance components depend on the genetic makeup of the population, with different genetic effects influencing trait inheritance. The additive component contributes to phenotype expression, while "missing heritability" may be due to epistasis, phenotypic plasticity, or rare genetic variants [53]. Al-Naggar et al. [54] demonstrated the importance of additive components over non-additive effects since the general combining ability (GCA) exceeded the specific combining ability (SCA) for grain protein, oil, and starch content. To improve this type of character in materials with wide genetic variability, an effective approach is often the recurrent selection of half-sib families [55]. Thus, a significant additive gene action effect in a specific population suggests that early-generation selection may result in the development of transgressive homozygous lines [56].

3. Materials and Methods

3.1. Genetic Material and Experimental Design

We used three commercial hybrids ("AX882" and "P1815" yellow-soft grain; "P2089" yellow-very soft grain), six inbred lines obtained by self-fertilization in inbreeding cycles at the Campo Escuela of the Facultad de Ciencias Agropecuarias, UNC ("B4" and "BL04" yellow-soft grain; "BCOT", "BulkASC", "C4B", and "CTM06" yellow-hard grain), and eight open-pollinated maize varieties developed ("BlancoM" white-soft grain; "C6006" orange-hard grain; "C8008" white-intermediate grain; "C900", "C980", "C990", "CandelariaINTA", and "LealesINTA" orange-intermediate grain) by selection for adaptation to the semi-arid central zone of Argentina. The genotypes were arranged in a randomized complete block design (RCBD) with two replications in experimental plots in the Campo Escuela (FCA, UNC) located in the semi-arid central region of Argentina (31°29′ S; 64°00′ W). The sowing dates were established in mid-December during the 2017–2018, 2018–2019, 2019–2020, and 2020–2021 seasons (Table 4). Conventional tillage (CT) was used for the first two seasons and no-till (NT) was used for the rest of the seasons. The experiments were conducted under rainfed conditions. The soil is an Entic Haplustoll and the superficial horizon presents a silt-loam texture, slightly acid to neutral, and well-supplied organic matter and is well-drained. Also, it does not present impediments that limit the crop growth.

Table 4. Average temperatures and cumulative rainfall during the critical period.

Year	RCc (mm)	RCcp (mm)	T° Minimumcp	T° Meancp	T° Maximumcp
2018	392	64.25 ± 12.14 a	15.36 ± 0.65 b	22.5 ± 0.58 c	30.26 ± 0.51 c
2019	557	109.98 ± 14.33 b	16.49 ± 0.24 d	22.11 ± 0.53 b	28.62 ± 0.71 b
2020	462	71.36 ± 42.54 a	16.07 ± 0.18 c	22.84 ± 0.23 d	30.29 ± 0.44 c
2021	588	96.16 ± 25.08 b	14.85 ± 1.03 a	20.58 ± 1.1 a	27.26 ± 1.23 a

Average values (±standard deviation) with different letters on each column for RCcp, T° Minimumcp, T° Meancp, and T° Maximumcp indicate statistical differences (Tukey's HSD test, $p \leq 0.05$). RCc, cumulative rainfall during the crop cycle; RCcp, cumulative rainfall in the critical period; cp, critical period.

3.2. *Agronomical Traits*

Agronomical traits included the following: plant height (Plh), recorded from the base of the stem to the apical end of the extended panicle (m); height of insertion of the main ear (He), recorded from the base of the stem to the node of the upper ear (most apical position) (m); stem diameter (Sd), recorded at the mean height of the first internode from the distal end of the stem (cm); ear length (El), recorded from distal to apical through the middle part of the spike (cm); ear diameter (Ed), measured in the middle part of the ear (cm); number of rows (Nr), recorded in the middle part of the ear; number of grains per row (Ngr), recorded from distal to apical in randomly selected rows; yield (Y), estimated using the production of each experimental plot and extrapolated to the area equivalent to one hectare (10,000 m^2). The yield results were adjusted to 14% humidity and expressed in quintals per hectare (q/ha). The traits were recorded on randomly selected plants until they reached 50% of the total plants in each plot. Thermal time to flowering (TT) was estimated from a base temperature of 10 °C. Climate data were collected during the crop critical period (anthesis ± 15 days) and included the minimum, average, and maximum temperatures as well as accumulated precipitation. These parameters were obtained from an agrometeorological station (Omixom SRL, Córdoba, Argentina) installed in this establishment.

3.3. *Grain Physical Properties*

The weight of 1000 grains (W1000) was calculated using an analytical balance [57]. The test weight (TW) was estimated using a Schopper balance of 250 mL total volume with a piston for air displacement [58]. The hardness of the grain was determined indirectly by the flotation index (FI) with a sodium nitrate solution (density (q) 1.250 ± 0.001) [59].

3.4. *Flour Chemical Traits*

Whole grain samples were ground using a Cyclotec CT193 cyclone mill (Foss, Suzhou, China) with a mesh aperture of 1 mm. The proximate composition was determined through proteins, lipids, and ash according to Methods 46-10.01, 30-25.01, and 08-01.01, respectively (AACC-approved methods of analysis, 2010). The carbohydrate content was estimated by the difference between the sum of the other components (proteins, lipids, and ashes). All determinations were expressed in g per 100 g of sample on a dry basis and were performed in duplicate.

3.5. *Statistical Analysis*

The InfoStat/Professional 2022 software (Facultad de Ciencias Agropecuarias, Universidad Nacional de Córdoba) and R 3.6.3 software (R Foundation for Statistical Computing, Vienna, Austria) were used. The data were examined using analysis of variance (ANOVA), with a significance level of 0.05. In addition, an LSD Fisher test for environmental parameters and a Tukey's HSD mean comparison test for agronomic, grain physical, and flour chemical traits, considering all harvest campaigns, were carried out [60,61]. Separate analyses were carried out according to the type of genetic material (OPV, L, and H). To appreciate the effects of the different sources of variation on the total variability, the variance components were estimated using Mixed Linear Models (MLMs) [62,63] using the restricted maximum likelihood estimator (REML) [64,65]. Thus, the MLM model_X_REML < −lmer(X~1 + (1 | Genotype) + (1 | Environment) + (1 | Genotype: Environment) with the lme4 statistical package version 1.1–35.5 was used, where X indicates the trait under study. The components of genotype, environment, and their interaction were considered random effects. Regarding the environmental factor, the crossing between the year and the tillage method used (conventional tillage and direct sowing), was considered. In this way, it was possible to obtain the magnitude of each variance (genotypic, environmental, and genotype–environment interaction) calculated as a percentage of the total variation. The relationships between the analyzed variables were determined using the Pearson correlation test, with a significance level of $p \leq 0.05$ and 0.01.

4. Conclusions

The intrinsic characteristics of the plant are essential to determine milling quality attributes such as grain hardness (measured indirectly by flotation index) and test weight. In addition, specific environmental factors from the semi-arid areas, such as rainfall and temperature, have a significant effect on grain composition in open-pollinated varieties, inbred lines, and hybrids. In particular, protein and lipid contents tend to decrease with higher temperatures during the critical growth period, while carbohydrate contents increase.

Inbred lines show significant genotype contributions to flotation index, protein, and carbohydrate content, which distinguishes them from the other two germplasm types. Identifying stability and sensitivity of specific traits—such as grain physical aspects and chemical composition—is crucial when selecting materials for potential improvement in variety development, targeting areas with stress conditions. This allows us to identify production environments where desired traits are optimally expressed.

The yield of the evaluated inbred lines was associated with a higher test weight and a lower grain weight and flotation index. Yield was also associated with higher lipid and lower carbohydrate contents. Considering the importance of ensuring food security in the current context of climate change, it is necessary to develop new crop varieties capable of producing high-quality raw materials in sufficient quantities for the industry. In this sense, the results of this study have identified specific genotypes, such as the inbred lines BL04 and B4, which are well-adapted to semi-arid conditions and have the potential to produce hybrids with high carbohydrate contents. In addition, inbred lines such as C4B, BulkASC, and BCOT can be used to produce hybrids with good protein contents.

Author Contributions: N.F.B.: data curation, formal analysis, investigation, methodology, resources, software, visualization, roles/writing—original draft, writing—review and editing. M.E.S.: investigation, methodology, supervision, validation, visualization, writing—review and editing. C.D.M.: data curation, formal analysis, software. C.A.B.: investigation, resources, supervision, visualization, writing—review and editing. A.E.L.: funding acquisition, investigation, methodology, project administration, resources, supervision, validation, visualization, writing—review and editing. All authors have read and agreed to the published version of the manuscript.

Funding: This work was funded by the National Agency for the Promotion of Research, Technological Development, and Innovation, Fund for Scientific and Technological Research (FONCyT), Ministry of Science and Technology of Argentina, N° 1019.

Data Availability Statement: The data presented in this study are available on request from the corresponding author.

Acknowledgments: The authors wish to thank the School of Agricultural Sciences—National University of Córdoba (FCA, UNC) and the Córdoba Food Science and Technology Institute (ICyTAC) for their technical assistance and provision of essential equipment for the preparation of this work.

Conflicts of Interest: The authors declared no potential conflicts of interest concerning the research, authorship, and/or publication of this article.

References

1. Chaudhary, D.P.; Kumar, S.; Yadav, O.P. Nutritive Value of Maize: Improvements, Applications and Constraints. In *Maize: Nutrition Dynamics and Novel Uses*; Springer: Berlin/Heidelberg, Germany, 2013; pp. 3–17.
2. Barreto, C.A.V.; das Graças Dias, K.O.; de Sousa, I.C.; Azevedo, C.F.; Nascimento, A.C.C.; Guimarães, L.J.M.; Guimarães, C.T.; Pastina, M.M.; Nascimento, M. Genomic Prediction in Multi-Environment Trials in Maize Using Statistical and Machine Learning Methods. *Sci. Rep.* **2024**, *14*, 1062. [CrossRef]
3. Koç, İ. Changes That May Occur in Temperature, Rain, and Climate Types Due to Global Climate Change: The Example of Düzce. *Turk. J. Agric.-Food Sci. Technol.* **2021**, *9*, 1545–1554. [CrossRef]
4. Trenberth, K.E. Changes in Precipitation with Climate Change. *Clim. Res.* **2011**, *47*, 123–138. [CrossRef]
5. Taylor, R.G.; Scanlon, B.; Döll, P.; Rodell, M.; Van Beek, R.; Wada, Y.; Longuevergne, L.; Leblanc, M.; Famiglietti, J.S.; Edmunds, M. Ground Water and Climate Change. *Nat. Clim. Chang.* **2013**, *3*, 322–329. [CrossRef]
6. Lee, C.-C.; Zeng, M.; Luo, K. How Does Climate Change Affect Food Security? Evidence from China. *Environ. Impact Assess. Rev.* **2024**, *104*, 107324. [CrossRef]
7. TAO, Z.; WANG, D.; CHANG, X.; WANG, Y.; YANG, Y.; ZHAO, G. Effects of Zinc Fertilizer and Short-Term High Temperature Stress on Wheat Grain Production and Wheat Flour Proteins. *J. Integr. Agric.* **2018**, *17*, 1979–1990. [CrossRef]
8. He, F.; Thiele, B.; Santhiraraja-Abresch, S.; Watt, M.; Kraska, T.; Ulbrich, A.; Kuhn, A.J. Effects of Root Temperature on the Plant Growth and Food Quality of Chinese Broccoli (*Brassica oleracea* Var. *Alboglabra bailey*). *Agronomy* **2020**, *10*, 702. [CrossRef]
9. Kusainova, A.A.; Mezentseva, O.V.; Tusupbekov, Z.A. Influence of Precipitation Variability and Temperature Conditions on the Yield of Grain Crops in Northern Kazakhstan. In *IOP Conference Series: Earth and Environmental Science*; IOP Publishing: Bristol, UK, 2020; Volume 548, p. 042026.
10. McDonald, A.J.; Keil, A.; Srivastava, A.; Craufurd, P.; Kishore, A.; Kumar, V.; Paudel, G.; Singh, S.; Singh, A.K.; Sohane, R.K. Time Management Governs Climate Resilience and Productivity in the Coupled Rice–Wheat Cropping Systems of Eastern India. *Nat. Food* **2022**, *3*, 542–551. [CrossRef]
11. Wang, X.; Wang, S.; Folberth, C.; Skalsky, R.; Li, H.; Liu, Y.; Balkovic, J. Limiting Global Warming to 2 °C Benefits Building Climate Resilience in Rice-Wheat Systems in India through Crop Calendar Management. *Agric. Syst.* **2024**, *213*, 103806. [CrossRef]
12. Biasutti, C.A.; Balzarini, M.B. Predicción Del Rendimiento de Híbridos de Maíz (*Zea mays* L.) En Ambientes de Siembra Tardía. *BAG J. Basic Appl. Genet.* **2017**, *28*, 19–26.
13. Ndhlela, T.; Herselman, L.; Magorokosho, C.; Setimela, P.; Mutimaamba, C.; Labuschagne, M. Genotype× Environment Interaction of Maize Grain Yield Using AMMI Biplots. *Crop Sci.* **2014**, *54*, 1992–1999. [CrossRef]
14. Katsenios, N.; Sparangis, P.; Chanioti, S.; Giannoglou, M.; Leonidakis, D.; Christopoulos, M.V.; Katsaros, G.; Efthimiadou, A. Genotype× Environment Interaction of Yield and Grain Quality Traits of Maize Hybrids in Greece. *Agronomy* **2021**, *11*, 357. [CrossRef]
15. Sharma, S.; Carena, M.J. Grain Quality in Maize (*Zea mays* L.): Breeding Implications for Short-Season Drought Environments. *Euphytica* **2016**, *212*, 247–260. [CrossRef]
16. Huang, C.; Duiker, S.W.; Deng, L.; Fang, C.; Zeng, W. Influence of Precipitation on Maize Yield in the Eastern United States. *Sustainability* **2015**, *7*, 5996–6010. [CrossRef]
17. Kothiyal, S.; Kaur, J. A Critical Analysis of the Effect of Projected Temperature and Rainfall for Differential Sowing of Maize Cultivars under RCP 4.5 and RCP 6.0 Scenarios for Punjab. *Theor. Appl. Climatol.* **2023**, *151*, 329–354. [CrossRef]
18. Bolsa de Cereales de Córdoba. La Agroindustria Cordobesa 2020–2021: Un Diagnóstico Para Mejorar Las Decisiones. Available online: https://www.bccba.org.ar/wp-content/uploads/2022/02/Anuario-DI-2021.pdf (accessed on 3 July 2024).
19. Sibanda, M.; Mushunje, A.; Mutengwa, C.S. An Evaluation on the Profitability of Growing Improved Maize Open Pollinated Varieties in the Eastern Cape Province, South Africa. *J. Dev. Agric. Econ.* **2016**, *8*, 1–13.
20. Kljak, K.; Novaković, K.; Zurak, D.; Jareš, M.; Pamić, S.; Duvnjak, M.; Grbeša, D. Physical properties of kernels from modern maize hybrids used in Croatia. *J. Cent. Eur. Agric.* **2020**, *21*, 543–553. [CrossRef]
21. Philippeau, C.; Landry, J.; Michalet-Doreau, B. Influence of the Protein Distribution of Maize Endosperm on Ruminal Starch Degradability. *J. Sci. Food Agric.* **2000**, *80*, 404–408. [CrossRef]
22. Nguma, E.; Murayama, D.; Munthali, C.; Onishi, K.; Mori, M.; Tani, M.; Palta, J.P.; Koaze, H.; Aiuchi, D. Effect of Kernel Type on Hardness and Interrelationship with Endosperm Chemical Components of Malawian Local Maize (*Zea mays* L.) Varieties during Storage. *Afr. J. Agric. Res.* **2020**, *16*, 1449–1457.
23. Cabañas-Ojeda, J.A.; Mejia-Abaunza, N.J.J.; Lozano-Cruz, P.A.; Aragão-Netto, V.; Brown, S.; Rubio, A.; Fahrenholz, A.; Oviedo-Rondón, E.O. Corn Kernel Hardness and Drying Temperature Affect Particle Size Post-Hammer-Milling and Pellet Quality in Broiler and Swine Diets. *Anim. Feed Sci. Technol.* **2023**, *304*, 115744. [CrossRef]
24. Nemati, A.; Sedghi, M.; Sharifi, R.S.; Seiedi, M.N. Investigation of Correlation between Traits and Path Analysis of Corn (*Zea mays* L.) Grain Yield at the Climate of Ardabil Region (Northwest Iran). *Not. Bot. Horti Agrobot. Cluj-Napoca* **2009**, *37*, 194–198.
25. Jamshidian, P.; Golparvar, A.R.; Naderi, M.R.; Darkhal, H. Phenotypic Correlations and Path Analysis between Ear Yield and Other Associated Characters in Corn Hybrids (*Zea mays* L.). *Int. J. Farming Allied Sci.* **2013**, *2*, 1273–1276.
26. Cerrudo, A.; Martinez, D.; Izquierdo, N.G.; Cirilo, A.G.; Laserna, M.P.; Reinoso, L.; Valentinuz, O.; Balbi, C.; Andrade, F.H. Environment, Management, and Genetic Contributions to Maize Kernel Hardness and Grain Yield. *Crop Sci.* **2017**, *57*, 2788–2798. [CrossRef]

27. Ciampitti, I.A.; Vyn, T.J. Physiological Perspectives of Changes over Time in Maize Yield Dependency on Nitrogen Uptake and Associated Nitrogen Efficiencies: A Review. *Field Crops Res.* **2012**, *133*, 48–67. [CrossRef]
28. Šeremešić, M.S.M.; Radosavljević, M.M.; Srdić, J.Ž.; Tomičić, Z.M.; Đuragić, O.M. Physical Traits and Nutritional Quality of Selected Serbian Maize Genotypes Differing in Kernel Hardness and Colour. *Food Feed Res.* **2019**, *46*, 51–60. [CrossRef]
29. Amegbor, I.K.; van Biljon, A.; Shargie, N.; Tarekegne, A.; Labuschagne, M.T. Heritability and Associations among Grain Yield and Quality Traits in Quality Protein Maize (QPM) and Non-QPM Hybrids. *Plants* **2022**, *11*, 713. [CrossRef]
30. Ramana Reddy, Y.; Ravi, D.; Ramakrishna Reddy, C.; Prasad, K.V.S.V.; Zaidi, P.H.; Vinayan, M.T.; Blümmel, M. A Note on the Correlations between Maize Grain and Maize Stover Quantitative and Qualitative Traits and the Implications for Whole Maize Plant Optimization. *Field Crops Res.* **2013**, *153*, 63–69. [CrossRef]
31. Werle, A.J.K.; Ferreira, F.R.A.; Pinto, R.J.B.; Mangolin, C.A.; Scapim, C.A.; Gonçalves, L.S.A. Diallel Analysis of Maize Inbred Lines for Grain Yield, Oil and Protein Content. *Crop Breed. Appl. Biotechnol.* **2014**, *14*, 23–28. [CrossRef]
32. Butts-Wilmsmeyer, C.J.; Seebauer, J.R.; Singleton, L.; Below, F.E. Weather during Key Growth Stages Explains Grain Quality and Yield of Maize. *Agronomy* **2019**, *9*, 16. [CrossRef]
33. Chen, G.-X.; Zhen, S.-M.; Liu, Y.-L.; Yan, X.; Zhang, M.; Yan, Y.-M. In Vivo Phosphoproteome Characterization Reveals Key Starch Granule-Binding Phosphoproteins Involved in Wheat Water-Deficit Response. *BMC Plant Biol.* **2017**, *17*, 168. [CrossRef]
34. Scrob, S.; Muste, S.; Has, I.; Muresan, C.; Socaci, S.; Fărcas, A. The Biochemical Composition and Correlation Estimates for Grain Quality in Maize. *J. Agroaliment. Process. Technol.* **2014**, *20*, 150–155.
35. Jahangirlou, M.R.; Akbari, G.A.; Alahdadi, I.; Soufizadeh, S.; Parsons, D. Grain Quality of Maize Cultivars as a Function of Planting Dates, Irrigation and Nitrogen Stress: A Case Study from Semiarid Conditions of Iran. *Agriculture* **2020**, *11*, 11. [CrossRef]
36. Song, K.; Kim, H.C.; Shin, S.; Kim, K.-H.; Moon, J.-C.; Kim, J.Y.; Lee, B.-M. Transcriptome Analysis of Flowering Time Genes under Drought Stress in Maize Leaves. *Front. Plant Sci.* **2017**, *8*, 267. [CrossRef] [PubMed]
37. Franks, S.J.; Sim, S.; Weis, A.E. Rapid Evolution of Flowering Time by an Annual Plant in Response to a Climate Fluctuation. *Proc. Natl. Acad. Sci. USA* **2007**, *104*, 1278–1282. [CrossRef] [PubMed]
38. Narváez-González, E.D.; Figueroa-Cárdenas, J.D.D.; Taba, S.; Sánchez, F.R. Kernel Microstructure of Latin American Races of Maize and Their Thermal and Rheological Properties. *Cereal Chem.* **2006**, *83*, 605–610. [CrossRef]
39. Martínez, R.D.; Cirilo, A.G.; Cerrudo, A.; Andrade, F.H.; Reinoso, L.; Valentinuz, O.R.; Balbi, C.N.; Izquierdo, N.G. Changes of Starch Composition by Postflowering Environmental Conditions in Kernels of Maize Hybrids with Different Endosperm Hardness. *Eur. J. Agron.* **2017**, *86*, 71–77. [CrossRef]
40. Beckles, D.M.; Thitisaksakul, M. How Environmental Stress Affects Starch Composition and Functionality in Cereal Endosperm. *Starch-Stärke* **2014**, *66*, 58–71. [CrossRef]
41. Caballero-Rothar, N.N.; Borrás, L.; Gerde, J.A. Physical and Chemical Kernel Traits Affect Starch Digestibility and Glycemic Index of Cooked Maize Flours. *Food Chem.* **2022**, *369*, 130953. [CrossRef]
42. Li, H.; Liu, Y.; Cheng, R.; Sun, X.; Wang, Y.; Tang, J.; Liu, Z. Inheritance Effect of Protein Content in Maize Kernels and Its Relation to Yield. *Acta Agron. Sin.* **2009**, *35*, 755–760. [CrossRef]
43. Mansilla, P.S.; Bongianino, N.F.; Nazar, M.C.; Pérez, G.T. Agronomic and Chemical Description of Open-Pollinated Varieties of Opaque-2 and Purple Maize (*Zea mays* L.) Adapted to Semiarid Region of Argentina. *Genet. Resour. Crop Evol.* **2021**, *68*, 2351–2366. [CrossRef]
44. Fang, H.; Fu, X.; Ge, H.; Zhang, A.; Shan, T.; Wang, Y.; Li, P.; Wang, B. Genetic Basis of Maize Kernel Oil-Related Traits Revealed by High-Density SNP Markers in a Recombinant Inbred Line Population. *BMC Plant Biol.* **2021**, *21*, 344. [CrossRef]
45. Mason, S.C.; D'croz-Mason, N.E. Agronomic Practices Influence Maize Grain Quality. *J. Crop Prod.* **2002**, *5*, 75–91. [CrossRef]
46. Ndlovu, N.; Spillane, C.; McKeown, P.C.; Cairns, J.E.; Das, B.; Gowda, M. Genome-Wide Association Studies of Grain Yield and Quality Traits under Optimum and Low-Nitrogen Stress in Tropical Maize (*Zea mays* L.). *Theor. Appl. Genet.* **2022**, *135*, 4351–4370. [CrossRef]
47. Ngaboyisonga, C.; Njoroge, K. Quality Protein Maize under Low-Nitrogen and Drought: Genotype by Environment Interaction for Grain and Protein Qualities. *Agric. J.* **2014**, *9*, 68–76.
48. Letchworth, M.B.; Lambert, R.J. Pollen Parent Effects on Oil, Protein, and Starch Concentration in Maize Kernels. *Crop Sci.* **1998**, *38*, 363–367. [CrossRef]
49. Tanaka, W.; Mantese, A.I.; Maddonni, G.A. Pollen Source Effects on Growth of Kernel Structures and Embryo Chemical Compounds in Maize. *Ann. Bot.* **2009**, *104*, 325–334. [CrossRef] [PubMed]
50. Li, H.; Fernie, A.R.; Yang, X. Using Systems Metabolic Engineering Strategies for High-Oil Maize Breeding. *Curr. Opin. Biotechnol.* **2023**, *79*, 102847. [CrossRef] [PubMed]
51. Ertiro, B.T.; Twumasi-Afriyie, S.; Blümmel, M.; Friesen, D.; Negera, D.; Worku, M.; Abakemal, D.; Kitenge, K. Genetic Variability of Maize Stover Quality and the Potential for Genetic Improvement of Fodder Value. *Field Crops Res.* **2013**, *153*, 79–85. [CrossRef]
52. Özdemir, E.; Sade, B. Genetic Analysis of Some Quality Traits in Maize. *Adnan Menderes Üniv. Ziraat Fak. Derg.* **2019**, *16*, 193–199. [CrossRef]
53. Li, H.; Wang, M.; Li, W.; He, L.; Zhou, Y.; Zhu, J.; Che, R.; Warburton, M.L.; Yang, X.; Yan, J. Genetic Variants and Underlying Mechanisms Influencing Variance Heterogeneity in Maize. *Plant J.* **2020**, *103*, 1089–1102. [CrossRef] [PubMed]
54. Al-Naggar, A.; Atta, M.; Ahmed, M.; Younis, A. Heterosis and Combining Ability of Maize (*Zea mays* L.) Grain Protein, Oil and Starch Content and Yield as Affected by Water Stress. *Arch. Curr. Res. Int.* **2016**, *4*, ACRI.27508. [CrossRef]

55. Njeri, S.G.; Makumbi, D.; Warburton, M.L.; Diallo, A.; Jumbo, M.B.; Chemining'wa, G. Genetic Analysis of Tropical Quality Protein Maize (*Zea mays* L.) Germplasm. *Euphytica* **2017**, *213*, 261. [CrossRef]
56. Cyplik, A.; Sobiech, A.; Tomkowiak, A.; Bocianowski, J. Genetic Parameters for Selected Traits of Inbred Lines of Maize (*Zea mays* L.). *Appl. Sci.* **2022**, *12*, 6961. [CrossRef]
57. Williams, P.J.; Kucheryavskiy, S. Classification of Maize Kernels Using NIR Hyperspectral Imaging. *Food Chem.* **2016**, *209*, 131–138. [CrossRef]
58. Ordóñez, M.R.; Gely, M.C.; Pagano, A.M. Estudio de Las Propiedades Físicas y de La Cinética de Secado de Granos de Maíz Colorado Duro. *Av. Cienc. Ing.* **2012**, *3*, 153–171.
59. Palacios-Rojas, N. Calidad Nutricional e Industrial de Maíz. In *Protocolos*; CIMMYT: Ciudad de México, Mexico, 2018; pp. 37–39.
60. Fisher, R.A. The Use of Multiple Measurements in Taxonomic Problems. *Ann. Eugen.* **1936**, *7*, 179–188. [CrossRef]
61. Nanda, A.; Mohapatra, B.B.; Mahapatra, A.P.K.; Mahapatra, A.P.K.; Mahapatra, A.P.K. Multiple comparison test by Tukey's honestly significant difference (HSD): Do the confident level control type I error. *Int. J. Stat.* **2021**, *6*, 59–65. [CrossRef]
62. Corbeil, R.R.; Searle, S.R. Restricted Maximum Likelihood (REML) Estimation of Variance Components in the Mixed Model. *Technometrics* **1976**, *18*, 31–38. [CrossRef]
63. Harville, D.A. Maximum Likelihood Approaches to Variance Component Estimation and to Related Problems. *J. Am. Stat. Assoc.* **1977**, *72*, 320–338. [CrossRef]
64. Schaeffer, L.R.; Wilton, J.W.; Thompson, R. Simultaneous Estimation of Variance and Covariance Components from Multitrait Mixed Model Equations. *Biometrics* **1978**, *34*, 199–208. [CrossRef]
65. Meyer, K. Maximum Likelihood Estimation of Variance Components for a Multivariate Mixed Model with Equal Design Matrices. *Biometrics* **1985**, *41*, 153–165. [CrossRef] [PubMed]

plants

MDPI

Systematic Review

A Meta-Analysis of the Effects of Increased Planting Density on Maize Yield in Northeast China

Junda Zhang [1], Xinyu Wang [1], Yuhao Li [1], Zikun Yu [1], Ruifang Zhang [2], Baozhong Yin [3] and Hongye Wang [1,*]

1 Cultivated Land Quality & Farmland Engineering Supervision and Protection Center, Ministry of Agriculture and Rural Affairs (MARA), Beijing 100020, China; jdsoil@163.com (J.Z.); wxy19951021@163.com (X.W.); talyh2016@163.com (Y.L.); yuzikun1@163.com (Z.Y.)

2 College of Land and Resources, Hebei Agricultural University, Baoding 071001, China; ruifangzhang2003@163.com

3 College of Plant Protection, Hebei Agricultural University, Baoding 071001, China; yinbaozhong@hebau.edu.cn

* Correspondence: wwwhhyy119@163.com

Abstract

Increasing maize planting density is considered as a potential strategy for enhancing grain yield in Northeast China; however, its yield-enhancing effects and underlying regulatory mechanisms remain inadequately characterized. Based on 508 paired observations from 42 publications, this meta-analysis quantified the effects of increased planting density on yield components and elucidated the regulatory roles of environmental and agronomic factors. Results demonstrated that increased density intensified plant competition, raising grains per ear by 43.7% and 100-kernel weight by 6.7%. The optimal regional planting density was determined to be 89,622.6 plants ha^{-1}, achieving a peak yield of 12,143 kg ha^{-1}—significantly exceeding current conventional densities (49,000–65,000 plants ha^{-1}) and highlighting substantial yield potential. Although nitrogen (N) management did not alter peak yield levels, it significantly increased the optimal density threshold (reaching 99,600 plants ha^{-1} under high N). Among environmental factors, mean annual precipitation (MAP) was the primary constraint: yield responses to increased density were negligible in arid regions (<400 mm); yield increase 4.0%, whereas optimal density (87,720 plants ha^{-1}) and peak yield (11,747.2 kg ha^{-1}) in high-rainfall regions ($\geq$400 mm) were significantly higher than in arid regions. Soil organic matter (SOM) (>20 g kg^{-1}) and optimal bulk density (BD) (1.25–1.40 g cm^{-3}) synergistically enhanced optimal density and peak yield. This study confirms the substantial, yet context-dependent, yield potential of increasing planting density in Northeast China, providing a science-based framework for region-specific optimization.

Keywords: maize; planting density; yield; meta-analysis; Northeast China; soil organic matter

1. Introduction

Improving food production to meet the needs of a growing global population remains a major challenge [1]. Maize (*Zea mays* L.) is one of the most widely cultivated cereal crops worldwide and serves as a staple food for over 4.5 billion people. Global maize demand is projected to double by 2050 to meet future food and feed requirements [2]. In China, where over 22% of the world's population depends on only 9% of global arable land, increasing maize productivity per unit area is crucial to ensure food security and sustainable agricultural development [3].

In addition to genetic improvement, optimized field management, and adequate water and nutrient supply, optimizing planting density is a key agronomic practice for achieving high maize yields. The Northeast China Plain (NECP) is the largest maize production base in China, accounting for over one-third of the national maize output [4]. However, planting densities in the region typically range from 49,000 to 65,000 plants ha^{-1}, which remains substantially lower than those in major producing regions like the United States (82,000–92,000 plants ha^{-1}) [5]. This suggests a large potential for yield improvement through enhanced planting density and optimized canopy structure in Northeast China.

Increasing planting density can raise maize yield primarily by boosting the number of ears per unit area and improving canopy light interception and dry matter accumulation. However, high plant density intensifies competition among plants for light, water, and nutrients, which can restrict leaf expansion, reduce photosynthetic capacity, accelerate leaf senescence [6], and limit grain filling, ultimately lowering yield per plant. Within an optimal range, the reduction in per-plant yield can be compensated by the increased plant population, resulting in higher yield per unit area [7]. Beyond this threshold, however, crowding stress leads to decreased biomass partitioning to kernels and a reduction in harvest index [8]. Consequently, previous studies in China have reported positive, neutral, or negative yield responses to increasing plant density, depending on climatic conditions, genotypes, and field management practices.

In Northeast China, where characterized by a cool temperate monsoon climate, short growing seasons, limited solar radiation, and variable soil fertility interact with density management to influence maize yield responses [9]. These complex interactions make it difficult to identify a single optimal planting density for regional maize production. Moreover, independent field studies have often produced inconsistent results even under similar density increments, reflecting the influence of confounding environmental and agronomic factors. This study provides a comprehensive, quantitative synthesis specific to Northeast China, uniquely evaluating the simultaneous moderating effects of climate, soil, and management factors on maize yield responses to planting density.

Meta-analysis provides a robust approach to synthesizing results across independent studies and quantifying the overall effects and moderators influencing the outcomes [10,11]. Therefore, the present study conducted a meta-analysis based on 508 paired observations from 42 published studies to comprehensively evaluate the effects of increased maize planting density on yield and related agronomic traits in Northeast China. The specific objectives were to:

(1) Quantify the impacts of increased planting density on maize yield and yield components; and (2) Identify the key agronomic, climatic, and edaphic factors regulating yield responses to density enhancement. The findings of this study aim to provide a scientific basis for optimizing maize planting density and improving yield potential under sustainable intensification in the Northeast China Plain.

2. Results

2.1. Overall Effects of Increased Planting Densities on Maize Yield

Increasing planting density significantly altered maize yield components (Figure 1). It reduced the number of grains per ear by 13.3% (95% CI: −14.8% to −11.7%) and the 100-kernel weight by 6.25% (95% CI: −6.2% to −5.0%). Conversely, it increased the number of ears per unit area by 43.7% (95% CI: 40.7% to 46.7%) and the grain yield by 6.7% (95% CI: 5.4% to 8.0%). Significant heterogeneity was detected across studies ($Q_M(df = 4) = 1618.45$, $p < 0.0001$), highlighting the substantial influence of moderating factors explored in subsequent sections.

https://doi.org/10.3390/plants15040544

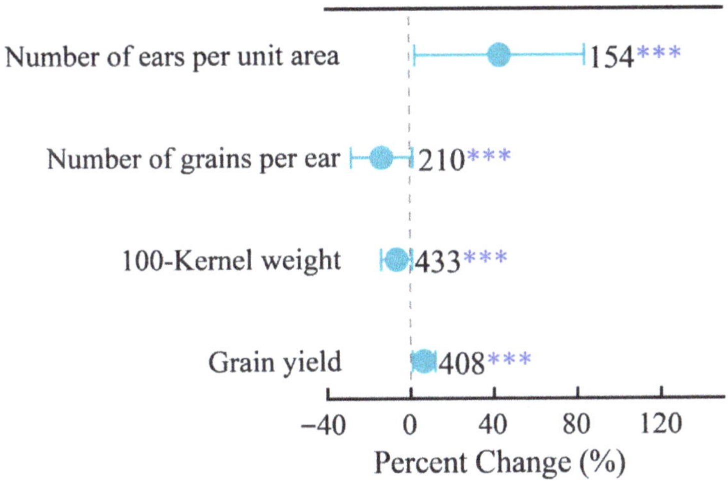

Figure 1. Effects of increased planting density on maize yield-related traits. Data are presented as percent changes relative to conventional planting density (CK). The closed circle indicates the mean percent change, and the horizontal line represents the 95% confidence interval. Asterisks denote significant difference from zero (*** $p < 0.001$).

2.2. *Optimal Planting Density and Maximum Grain Yield of Maize in Northeast China*

Quadratic regression analysis revealed that maize grain yield followed a parabolic relationship with planting density, initially increasing and then declining. The model projected a maximum yield of 12,143.0 kg·ha^{-1} at the optimal density of 89,622.6 plants·ha^{-1}. Notably, grain yield began to decline when density exceeded 12 × 10^4 plants·ha^{-1} (Figure 2).

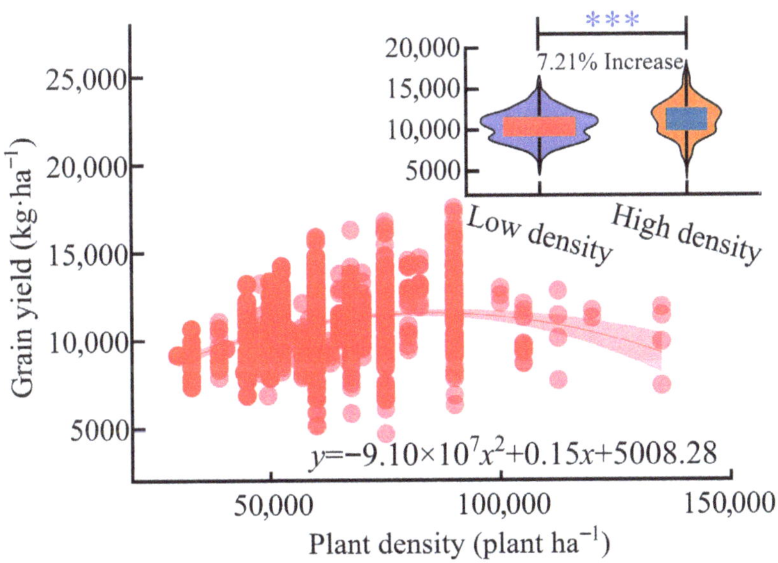

Figure 2. Relationship between planting density and maize grain yield in Northeast China. Scatter points represent field-observed grain yield under different planting densities (×10^4 plants ha^{-1}). (*** $p < 0.001$). The quadratic regression model shows that the red line represents the changes in yield under different planting densities; the box plot indicates the data of yield after low-density planting (blue) and the data of yield after high-density planting (orange).

https://doi.org/10.3390/plants15040544

Furthermore, box plot analysis confirmed significantly higher grain yields under higher-density regimes (>6 × 10^4 plants ha^{-1}) compared to lower-density groups (≤6 × 10^4 plants ha^{-1}), with an 18.5% increase in median yield.

2.3. Effects of Nitrogen Application Rates on Maize Yield Response to Increased Planting Density

The yield response to increased planting density was strongly moderated by nitrogen (N) application rates (Figure 3) ($Q_M(df = 4) = 247.8535$, $p < 0.0001$). Based on regional fertilization practices, N rates were categorized as low (<180 kg N ha^{-1}), medium (180–240 kg N ha^{-1}), and high (>240 kg N ha^{-1}) [5,8]. Subgroup analysis revealed a significant yield increase only under low N conditions (+7.6%; 95% CI, 5.1–10.0%; $p < 0.001$), with non-significant responses under medium and high N.

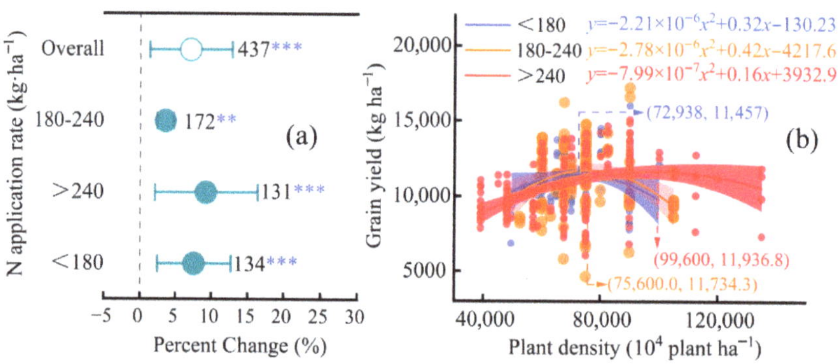

Figure 3. (**a**) Percent change in maize grain yield under increased planting density relative to conventional density, stratified by nitrogen (N) application rates: low (<180 kg·ha^{-1}), medium (180–240 kg ha^{-1}), high (>240 kg ha^{-1}), and overall (all rates combined). (**b**) Quadratic regression models illustrating the relationship between planting density (plants ha^{-1}) and maize grain yield (kg ha^{-1}) for low (blue), medium (orange), and high (red) N application rates. (** $p < 0.01$, *** $p < 0.001$).

Quadratic regression modeling further elucidated the density-yield dynamics under varying N inputs (Figure 3b). The optimum planting density (i.e., the density maximizing grain yield) increased with N availability: 72,398 plants ha^{-1} under low N, 75,600 plants ha^{-1} under medium N, and 99,600 plants ha^{-1} under high N. Notably, although the optimal density rose by approximately 38% from low to high N, the corresponding peak grain yields, the peak grain yields achieved at these optimal densities did not differ significantly among the three N-rate categories.

2.4. Effects of Mean Annual Temperature (MAT) and Mean Annual Precipitation (MAP) on Maize Yield Response to Increased Planting Density

The yield response to increased planting density was significantly moderated by mean annual precipitation (MAP) but not by mean annual temperature (MAT) (Figure 4). Overall, increasing density significantly raised yield by 6.7% (95% CI, 5.5–7.9%; $Q_M(df = 3) = 255.2571$, $p < 0.0001$). When stratified by MAT, yield increases were significant in both low (<7 °C; +8.2%) and medium (7–14 °C; +5.0%) temperature groups, with no statistical difference between them ($p > 0.05$). In contrast, MAP had a pronounced effect: the yield response was negligible in arid regions (<400 mm; +4.0%, $p > 0.05$) but significant in regions with ≥400 mm precipitation (+7.0%; $p < 0.001$). Quadratic regression indicated that the optimal planting density and peak yield were substantially higher in the high-MAP region (87,720 plants ha^{-1} and 11,747 kg ha^{-1}, respectively) compared to the low-MAP region (57,143 plants ha^{-1} and 11,292 kg ha^{-1}) (Figure 4d).

https://doi.org/10.3390/plants15040544

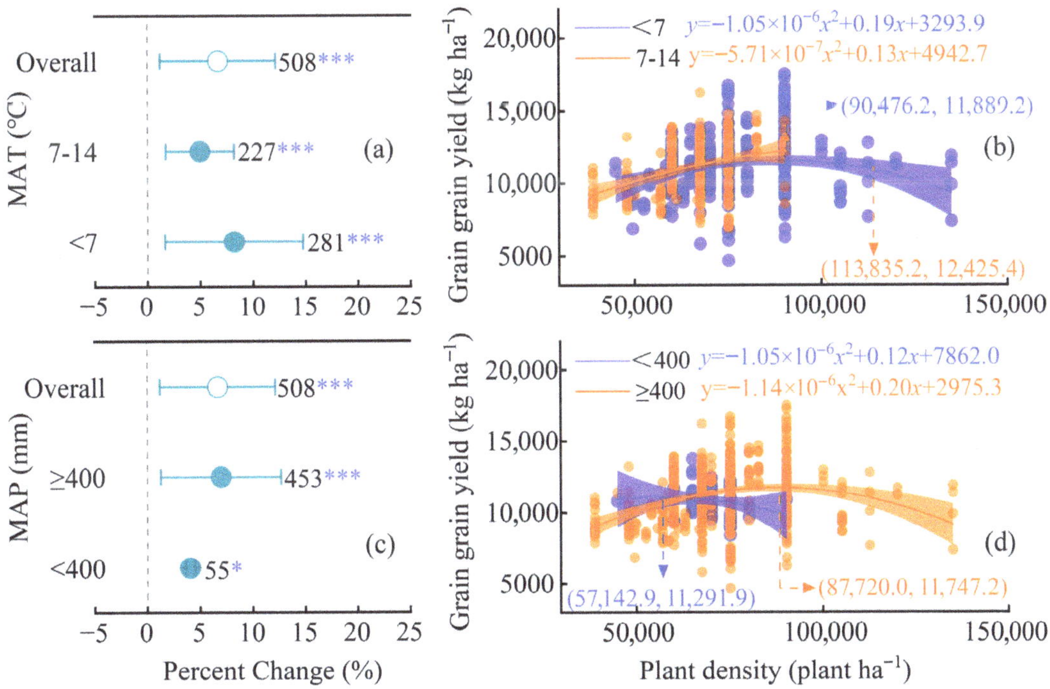

Figure 4. (**a**) Yield response to increased planting density across MAT groups. (**b**) Density–yield relationships for MAT groups; low MAT: <7 °C, blue circles; high MAT: 7–14 °C, orange circles. (**c**) Yield response across MAP groups. (**d**) Density–yield relationships for MAP groups; low MAP: <400 mm, blue circles; high MAT: ≥400 mm, orange circles. (* $p < 0.05$, *** $p < 0.001$).

2.5. Effects of Increasing the Planting Density Under Different Soil Conditions

All measured soil factors significantly moderated the density–yield relationship ($p < 0.001$; Figure 5). The yield increase was most pronounced under low total nitrogen (TN $\leq$ 0.8 g kg^{-1}), reaching 12.9% (95% CI: 6.8 to 19.3%), and was significantly higher than under high TN ($p < 0.001$). Quadratic models showed the highest optimal density (90,476 plants ha^{-1}) occurred at high TN, but the highest peak yield (13,226 kg ha^{-1}) at low TN (Figure 5a,b).

For soil organic matter (SOM), yield response was stronger in low-SOM soils (<10 g kg^{-1}), while optimal density and peak yield increased progressively with SOM content (Figure 5c,d). Although yield increased significantly across pH groups, the response was greater in acidic soils (pH $\leq$ 6.5), whereas optimal density was higher in neutral-to-alkaline soils (Figure 5e,f).

Critically, bulk density (BD) exhibited a clear optimum: yield response peaked within the range of 1.25–1.40 g cm^{-3} (Figure 5g,h).

2.6. Model-Averaged Relative Importance of Predictors in Explaining Maize Yield Responses to Increased Planting Density

Model-averaged analysis (Figure 6) identified mean annual precipitation (MAP), soil organic matter (SOM), and bulk density (BD) as the most influential factors moderating the yield response to increased planting density. Nitrogen application rate and mean annual temperature (MAT) played moderate roles, while total soil nitrogen (TN) and pH contributed minimally.

 https://doi.org/10.3390/plants15040544

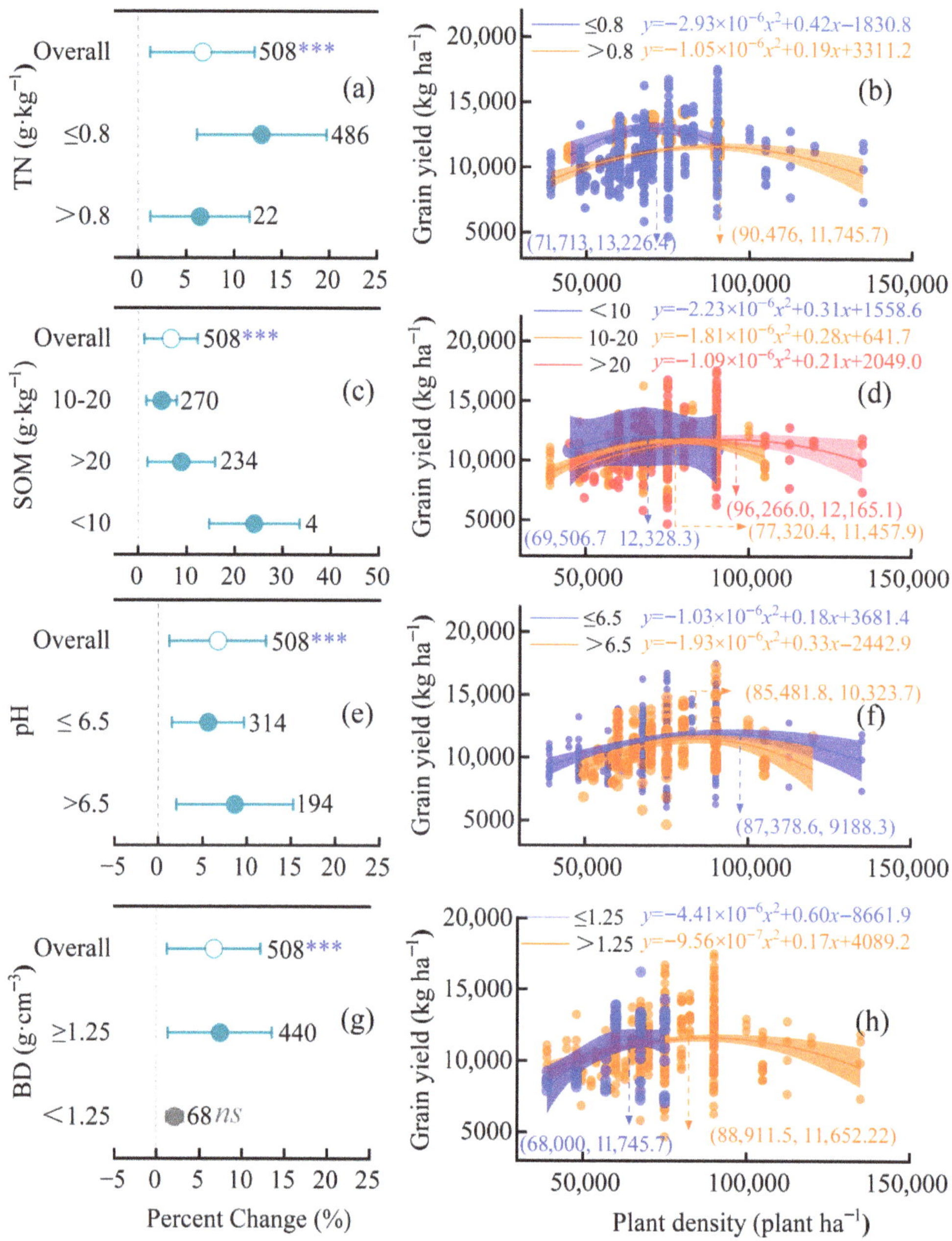

Figure 5. Effects of soil properties on maize yield response to increased planting density. (**a**,**c**,**e**,**g**) Percent yield change relative to conventional density, stratified by soil property: total nitrogen (TN), soil organic matter (SOM), pH, and bulk density (BD). *** $p < 0.001$; ns, not significant. (**b**,**d**,**f**,**h**) Quadratic regression of planting density versus grain yield. Data points and curves are color-coded by subgroup: (**b**) TN (low: $\leq$0.8 g·kg^{-1}, blue circles; high: >0.8 g·kg^{-1}, orange circles); (**d**) SOM (low: <10 g·kg^{-1}, blue circles; medium: 10–20 g·kg^{-1}, orange circles; high: >20 g·kg^{-1}, red circles;) (**f**) pH (low: $\leq$6.5, blue circles; high: >6.5, orange circles); (**h**) BD (low: $\leq$1.25 g·cm^{-3}, blue circles; high: >1.25 g·cm^{-3}, orange circles).

https://doi.org/10.3390/plants15040544

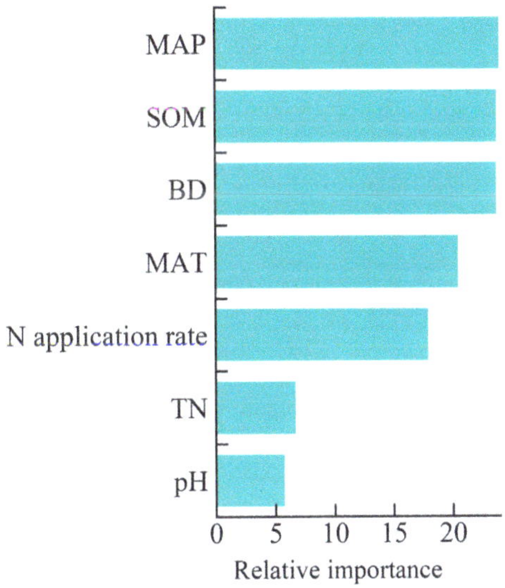

Figure 6. Relative importance of environmental (MAP, MAT), soil (SOM, BD, TN, pH), and management (N application) factors in explaining maize yield responses to increased planting density.

3. Discussion

Our meta-analysis, synthesizing data from 508 field observations, reveals a complex and heterogeneous relationship between planting density and maize yield in Northeast China. This substantial variability ($I^2 > 80\%$) is not a limitation but rather a key finding: it underscores that the yield response to densification is profoundly context-dependent. The primary objective of this discussion is to decipher this heterogeneity by quantifying and interpreting the moderating roles of key agronomic, climatic, and edaphic factors. Below, we explore the trade-offs and synergies that define the optimal density window across diverse production contexts.

3.1. *Trade-Offs Between Population-Level Gains and Individual Plant Constraints*

Our meta-analysis confirms that increasing planting density enhances maize yield at the population level (+5.3%) but imposes significant constraints on individual plant growth (Figure 1). The observed reduction in grains per ear (−13.3%) and 100-kernel weight (−6.3%) aligns with established physiological principles: intensified competition for light, water, and nutrients under high density suppresses photosynthetic capacity, leaf expansion, and dry matter partitioning to kernels. In densely planted stands, individual plants experience reduced radiation interception per leaf area, leading to incomplete kernel development and lower sink strength, which may be attributed to source limitation or altered assimilate partitioning during grain filling [12].

This trade-off—where yield gains at the population scale compensate for per-plant yield penalties—represents a fundamental feature of high-density cropping systems and helps explain why density optimization, rather than maximization, is critical for productivity [7,12,13]. Increased canopy closure improves radiation interception efficiency, particularly during the early reproductive stages, yet excessive crowding accelerates leaf senescence, increases self-shading, and suppresses lower canopy photosynthesis. The yield-density relationship therefore follows a quadratic trend: yield initially rises with density due to improved light capture and resource use, then plateaus or declines when resource competition outweighs collective gains.

https://doi.org/10.3390/plants15040544

In Northeast China, our calculated optimal density of 89,622.6 plants ha^{-1} (yielding 12,143.0 kg ha^{-1}; Figure 2) represents a 37–83% increase over current regional practices (49,000–65,000 plants ha^{-1}). This substantial gap highlights untapped potential for closing yield gaps in the region. The results also suggest that many existing density recommendations may be outdated, as they were established for older, less compact cultivars. Modern hybrids with improved canopy architecture and stress tolerance can maintain higher photosynthetic efficiency and harvest index under crowding, justifying an upward revision of density thresholds. Additionally, optimizing spatial uniformity-via precision planting or mechanical adjustment-can further balance intra-row competition and enhance per-area productivity.

3.2. Nitrogen Management: Regulating Density Thresholds Without Altering Yield Ceilings

Changes in yield components are not driven solely by planting density in isolation, but by its interaction with key agronomic and environmental factors. The negative effects of high density on individual plants can be significantly mitigated under optimal nitrogen and water management, as demonstrated by our subgroup analyses (Figures 3–5).

This study reveals a unique role of nitrogen (N) fertilization in high-density maize systems: N application rate significantly modulates optimal planting density but does not alter peak grain yield (Figure 3). Under low N inputs (<180 kg ha^{-1}), the optimal density was 72,398 plants ha^{-1}, while high N inputs (>240 kg ha^{-1}) increased optimal density by 38% to 99,600 plants ha^{-1}. Crucially, peak yields remained statistically invariant across low, medium, and high N rates (Figure 3b).

This finding indicates that N availability primarily determines the density capacity—the threshold of population tolerance to intraspecific competition—rather than the attainable yield ceiling, which remains constrained by light and hydrothermal resources [14,15]. Nitrogen promotes chlorophyll synthesis, prolongs leaf longevity, and enhances photosynthetic resilience under crowding, thereby allowing denser stands to sustain growth. However, once canopy photosynthesis reaches saturation, additional N cannot further raise yield ceilings, as grain filling becomes limited by radiation and temperature.

These results challenge the conventional paradigm that higher N inputs directly translate into yield gains. In high-density systems, N functions more as a "modulator" enabling structural adaptation and resource balance rather than as a "driver" of yield enhancement. From a management standpoint, this implies that synchronizing N availability with plant demand—through split applications or slow-release fertilizers—can stabilize yield while minimizing losses. Optimized N scheduling can also mitigate the risk of premature leaf senescence often observed in N-deficient, densely planted crops. Thus, strategic N management supports the sustainability of intensive maize systems without compromising environmental integrity.

3.3. Climate and Soil Mediators of Density Efficacy

Climatic regulation of maize yield responses to increased planting density exhibited marked heterogeneity, with mean annual precipitation (MAP) emerging as the primary limiting factor. In low-MAP regions (<400 mm), density increases yielded negligible yield gains (+4.0%, $p > 0.05$), whereas high-MAP areas ($\geq$400 mm) achieved significant yield improvements (+7.0%), supported a substantial surge in optimal planting density, and elevated peak grain yield (Figure 4c,d). This divergence underscores water's critical role in mitigating competition stress-ample precipitation ensures balanced water allocation under high-density conditions [16]. Sufficient water availability not only alleviates stomatal limitation but also stabilizes canopy temperature and facilitates nutrient uptake, leading to more synchronized ear development across plants.

 https://doi.org/10.3390/plants15040544

Conversely, mean annual temperature (MAT) exerted no significant influence: yield responses showed no statistical differences across low (<7 °C), medium (7–14 °C), and high (>14 °C) MAT groups (3.2–9.9%; $p > 0.05$; Figure 4a). This apparent temperature resilience likely reflects maize 's broad adaptability within the temperate monsoon climates of Northeast China. It should be noted that our use of MAT, rather than growing-season-specific metrics, may integrate both critical and non-critical thermal periods [17]. Therefore, while our analysis suggests MAT is not a primary moderator, future research employing finer temporal resolution could refine insights into temperature effects during key developmental stages. Furthermore, projected climate warming may extend the growing season and alter density optima, implying that future management strategies must integrate considerations of both cumulative heat units and water balance.

Soil properties exhibited hierarchical control over density responses, where soil organic matter (SOM) and bulk density (BD) were pivotal regulators. High-SOM soils (>20 g kg^{-1}) increased optimal density by 27.8% and reduced the maximum yield by only 1.3% compared to low-SOM soils (Figure 5d). SOM buffers high-density stress by enhancing soil structure, water retention, and nutrient supply capacity, thereby promoting sustained root activity during critical growth stages. Similarly, moderate BD (1.25–1.40 g cm^{-3}) optimized root-zone conditions, ensuring better root penetration and aeration. Yields under this ideal BD exceeded those in looser soils (<1.25 g cm^{-3}) by 71.1% (Figure 5h), suggesting that adequate compaction supports stable anchorage and root–soil contact for efficient water use.

In contrast, total nitrogen (TN) and pH played minimal roles. Model-averaged analysis ranked TN as the least important predictor (Figure 6), and although low-TN soils (≤0.80 g kg^{-1}) showed stronger relative yield responses (+12.9%), their absolute peak yields remained lower than those of medium/high-TN groups (Figure 5b). This confirms that native soil N cannot substitute for fertilizer N in regulating density thresholds, while pH's weak effect derives from its general suitability (6–8) across the region. Overall, soil quality and water availability jointly define the physiological buffer zone within which density intensification can be successful.

3.4. Strategic Intensification: Synergizing Density, Nitrogen, and Soil Health

To harness the yield potential of high-density maize systems in Northeast China, three interdependent strategies emerge. First, prioritize density escalation to $\geq 9.0 \times 10^4$ plants·ha^{-1} in high-MAP zones (>600 mm; e.g., Eastern Songnen Plain), while integrating water-saving techniques (drip irrigation, mulching) in arid areas (<400 mm) to overcome precipitation limitations [18–20]. Second, implement precision nitrogen management by aligning N inputs (180–240 kg ha^{-1}) with target planting densities (e.g., higher N for densities > 9.0×10^4 plants ha^{-1}) to mitigate environmental risks without compromising yield ceilings—a critical insight given N's role in expanding density capacity rather than elevating yield maxima [21,22]. Third, enhance soil health through residue retention and organic amendments to boost SOM (>20 g kg^{-1}), coupled with subsoiling to maintain optimal BD (1.25–1.40 g cm^{-3}), thereby creating resilient root environments for high-density stands.

Beyond these agronomic interventions, integrating digital farming tools-such as remote sensing, canopy imaging, and data-driven decision support-can further refine density–nutrient interactions across spatially variable fields. This precision intensification paradigm not only maximizes yield but also improves input efficiency and ecosystem resilience.

This integrated approach—density-driven intensification conditioned on water availability, precision-nutrient matching, and soil health investment—resolves the trade-off between population-level gains and individual plant constraints, unlocking sustainable yield

https://doi.org/10.3390/plants15040544

advances in NECP maize systems [23–25]. It offers a practical roadmap toward climate-smart, resource-efficient maize production aligned with regional sustainability goals.

4. Materials and Methods

4.1. Data Collection

A systematic literature search was conducted to evaluate the effects of increased planting density on maize (*Zea mays* L.) grain yield in Northeast China. Following the PRISMA (Preferred Reporting Items for Systematic Reviews and Meta-Analyses) guidelines (Figure 7), peer-reviewed publications from 1990 to June 2025 were retrieved from the Web of Science, Scopus, Google Scholar, and China National Knowledge Infrastructure (CNKI) databases. The search combined English and Chinese keywords, including "maize" OR "corn", "planting density" OR "plant population" OR "row spacing", "grain yield", and "Northeast China" OR "Heilongjiang" OR "Jilin" OR "Liaoning" OR "Songnen Plain" OR "Sanjiang Plain." The search was performed in the title, abstract, and keyword fields using Boolean operators e.g., TI = ("maize") AND AB = ("planting density") AND AB = ("yield"). Studies were included based on the following criteria: (1) field experiments conducted in Northeast China; (2) maize as the target crop, with at least two planting density treatments (one serving as the control representing local or conventional density, and the other representing increased density); (3) provision of at least one outcome variable related to grain yield; and (4) availability of sufficient statistical information such as standard deviation (SD), standard error (SE), or sample size (n). When multiple sites or years were reported in one study, each site × year combination was treated as an independent observation. Mean annual temperature (MAT) was used as a consistent climatic proxy because specific growing-season temperature metrics were not consistently reported across the compiled studies.

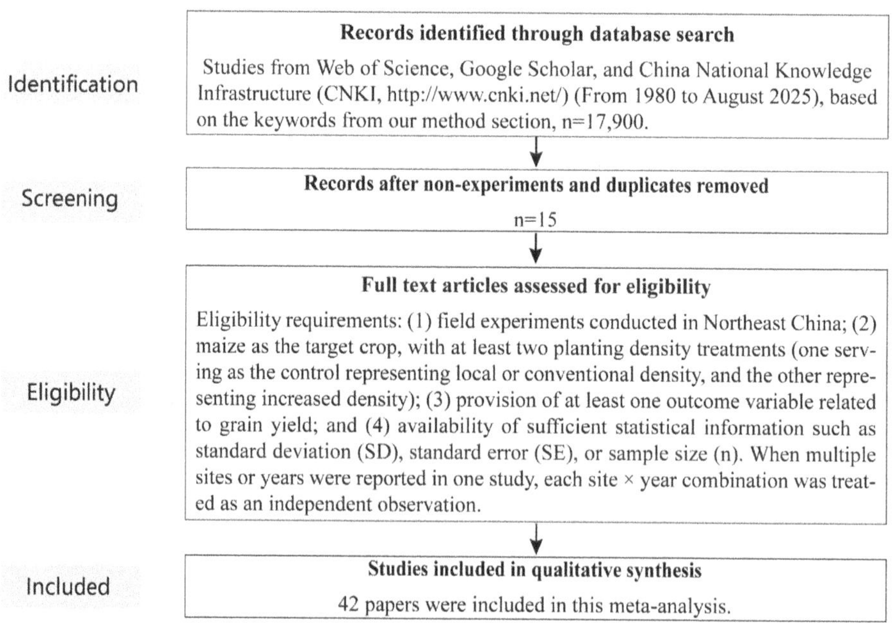

Figure 7. Preferred Reporting Items for Systematic Reviews and Meta-Analysis (PRISMA) flow chart illustrating the study selection procedure for the meta-analysis.

From each eligible publication, the following data were extracted:

https://doi.org/10.3390/plants15040544

(1) General information, including author name, publication year, and study location (latitude and longitude if available); (2) experimental details, including planting density (plants ha^{-1}) or row spacing (cm), cultivar type (compact or conventional), fertilizer N rate (kg ha^{-1}), and management practices (e.g., irrigation, tillage); (3) environmental variables, such as soil type, mean annual temperature (MAT), and mean annual precipitation (MAP); and (4) outcome variables, including maize grain yield and yield components (if reported).

All data were digitized and organized using a standardized data extraction template. For studies presenting data only in graphical format, numerical values were obtained using WebPlotDigitizer (version 3.4). A total of 42 eligible publications met the inclusion criteria, yielding 508 paired observations across Northeast China. The geographic distribution of the study sites is shown in Figure 8.

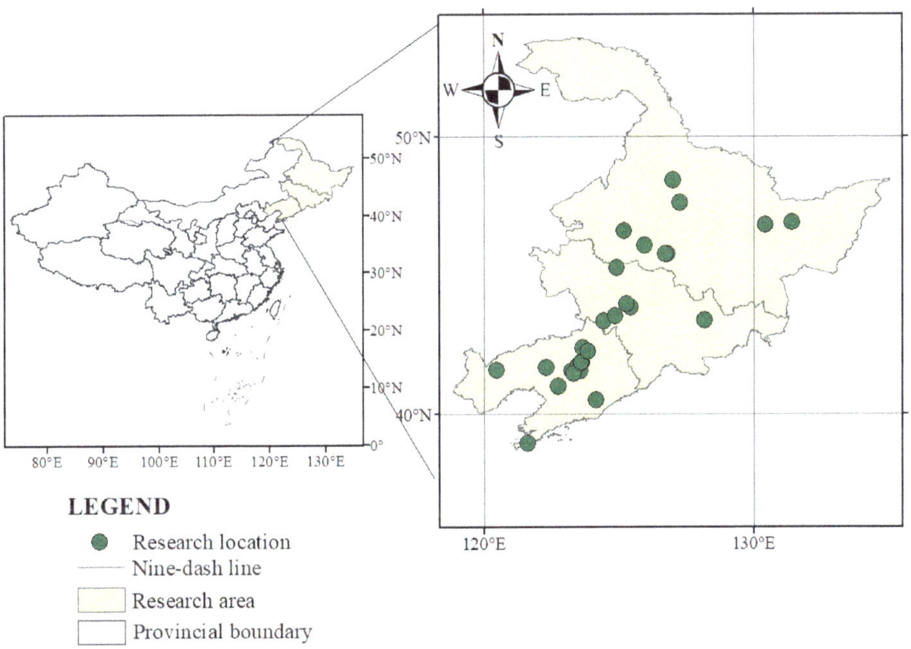

Figure 8. Geographic distribution of planting density experiments across 42 independent study sites included in the meta-analysis.

Mean annual temperature (MAT) was employed as the primary thermal metric due to the consistent reporting of site-level climate normals in the source literature, whereas detailed growing-season temperature data were largely unavailable. While MAT and growing season temperature are correlated in this region, we acknowledge that MAT may not fully capture temperature effects during critical phenological stages. This limitation is considered in the interpretation of the results.

4.2. Meta-Analysis Procedure

To quantify how planting density and its driving factors affect yield, we used the natural-log response ratio (*lnR*) as the effect-size metric—a choice widely adopted in ecological and agronomic meta-analyses.

$$lnR = ln(X_t/X_c) \quad (1)$$

https://doi.org/10.3390/plants15040544

Here, X_t and X_c denote the means of the treatment (increased planting density) and control (conventional density) groups, respectively.

The variance (v) of lnR for each observation was calculated as follows:

$$v = {SD_t}^2/n_t{X_t}^2 + {SD_c}^2/n_c{X_c}^2 \quad (2)$$

Here, SD_t, SD_c, n_t, and n_c represent the standard deviations and replicate numbers of the treatment and control groups, respectively.

The weighted effect sizes (lnR_{++}) were determined by the following equation:

$$lnR_{++} = \sum(lnR_i\ W_i)/\sum\ (W_i) \quad (3)$$

Here, lnR_i is the effect size of the i-th comparison and W_i is the corresponding weight, defined as follows:

$$W_i = 1/\left(v_i + \tau^2\right) \quad (4)$$

Here, v_i denotes the sampling variance of lnR_i, and τ^2 is the between-study variance, estimated using the restricted maximum likelihood (REML) method in the rma.mv function of the R package "metafor" (version 4.5.1), implemented in R version 4.5.1.

The standard error of the weighted mean and its 95% confidence interval (CI) were calculated as follows:

$$S_{lnR++} = \sqrt{1/\sum W_i} \quad (5)$$

$$95\%\ \mathrm{CI} = lnR_{++} \pm 1.96\ S_{lnR++} \quad (6)$$

For studies that did not report standard deviations (SDs), the impute_SD function in the 'metagear' package (R version 4.5.1) was employed to impute missing values [25]. To account for the non-independence of effect sizes arising from shared control groups, a variance–covariance matrix was constructed following the method of Lajeunesse [26].

A random-effects model was employed; this modeled variance at the study level, treatment level, and sampling error level. Parameter estimation was conducted using restricted maximum likelihood (REML). The 'rma' function in the 'metafor' package (R version 4.5.1) was used to estimate the weighted effect sizes (lnR_{++}) and their 95% confidence intervals (CIs). An effect was deemed statistically significant if the CI did not overlap with zero. Between-group differences were considered significant if the respective CIs did not overlap [27]. Heterogeneity among studies was assessed using the Q statistic (Q_t), with significance indicating variation in effect sizes potentially attributable to moderator variables [28]. Subgroup analyses were conducted for categorical moderators with at least ten observations or at least five observations from two or more independent studies [29]. Meta-regression analyses were conducted using the "rma()" function with the restricted maximum-likelihood estimator (REML) in the "metafor" package (R version 4.5.1) to examine the relationships between effect sizes and climatic variables, including mean annual precipitation (MAP) and mean annual temperature (MAT) [30,31].

To enhance interpretability, effect sizes were transformed into percentage changes using the following equation:

$$Percent\ change = \left(exp^{lnR} - 1\right) \times 100\% \quad (7)$$

Using the "randomForest" package in R software version 4.2.2, a random forest model was developed to quantify the relative importance of environmental, soil, and management factors in explaining maize yield responses to increased planting density. The model incorporated environmental variables (mean annual precipitation, MAP; mean annual

 https://doi.org/10.3390/plants15040544

temperature, MAT), soil properties (soil organic matter, SOM; bulk density, BD; total nitrogen, TN; and pH), and management factors (nitrogen application rate). Statistical significance of variable importance was evaluated using the "rfPermute" package, with significance accepted at $p < 0.05$, indicating robust model performance. The relative importance of each factor was ranked based on the percentage increase in mean squared error (MSE) when that variable was permuted, averaged over 500 iterations of the random forest model.

Heterogeneity and publication bias were assessed. The random-effects model indicated substantial heterogeneity among effect sizes ($I^2 = 82.97\%$, $\tau^2 = 0.0061$), which was anticipated given the context-dependent nature of density effects. The subgroup and meta-regression analyses were performed to elucidate the sources of this heterogeneity. Egger's regression test found no significant publication bias ($p = 0.15$).

5. Conclusions

This meta-analysis demonstrates that increasing planting density can substantially enhance maize yields in Northeast China, but with a clear optimum. Mean annual precipitation (MAP), soil organic matter (SOM), and bulk density (BD) were identified as the primary limiting factors. In contrast, nitrogen (N) management primarily regulates the tolerable plant density threshold rather than the maximum achievable yield. For regions with ≥400 mm annual precipitation, we recommend a target density of approximately 87,700 plants ha^{-1} coupled with an N application rate of 180–240 kg N ha^{-1}. Combine this with organic amendments and occasional deep tillage to lift soil organic matter above 20 g kg^{-1} and keep bulk density between 1.25 and 1.40 g cm^{-3}. Such an integrated package can unlock the region's untapped yield potential while keeping production sustainable.

Author Contributions: J.Z.: investigation, writing—original draft, project administration, methodology. X.W.: writing—review and editing, conceptualization. Y.L.: investigation, Formal analysis. Z.Y.: investigation, Formal analysis. R.Z.: investigation, Formal analysis. B.Y.: investigation, Formal analysis. H.W.: conceptualization, writing—review and editing. All authors have read and agreed to the published version of the manuscript.

Funding: This work was supported by the National Key R&D Program of China (2021YFD1500205).

Data Availability Statement: Data is contained within the article.

Conflicts of Interest: The authors declare no conflicts of interest.

References

1. FAO. *OECD-FAO Agricultural Outlook 2023–2032*; FAO: Rome, Italy, 2023. [CrossRef]
2. FAO. *World Food Outlook—Cereals Demand to 2050*; FAO: Rome, Italy, 2023.
3. Zhang, Y.; Zhang, G.; Zhai, J.; Cao, Y.; Xu, W.; Ming, B.; Xie, R.; Wang, K.; Li, S.; Xue, J.; et al. Post-Silking Nitrogen Topdressing Optimizes Nitrogen Accumulation and Enhances Yield in Densely Planted Maize. *Agronomy* **2026**, *16*, 26. [CrossRef]
4. Liu, H.-J.; Gao, Z.-Z.; Zhang, L.-W.; Liu, Y. Stomatal conductivity, canopy temperature and evapotranspiration of maize (*Zea mays* L.) to water stress in Northeast China. *Int. J. Agric. Biol. Eng.* **2021**, *14*, 112–119. [CrossRef]
5. Zhao, Y.-J.; Xing, S.; Zhang, Q.-S.; Zhang, F.-S.; Ma, W.-Q. Causes of maize density loss in farmers' fields in Northeast China. *J. Integr. Agric.* **2019**, *18*, 1680–1689. [CrossRef]
6. Liu, Y.P.; Wang, S.Y.; Li, M.L.; Zhang, F.; Wang, R.-W. The emergence of adaptive diversification from plant's light competition. *Chaos Solitons Fractals* **2021**, *152*, 111366. [CrossRef]
7. Zheng, Y.; Yue, Y.; Li, C.; Wang, Y.; Zhang, H.; Ren, H.; Gong, X.; Jiang, Y.; Qi, H. Revolutionizing Maize Crop Productivity: The Winning Combination of Zigzag Planting and Deep Nitrogen Fertilization for Maximum Yield through Root–Shoot Ratio Management. *Agronomy* **2023**, *13*, 1307. [CrossRef]
8. Zhang, G.; Cui, C.; Lv, Y.; Wang, X.; Wang, X.; Zhao, D.; Hu, F.; Wen, X.; Han, J.; Liao, Y. Is it necessary to increase the maize planting density in China? *Eur. J. Agron.* **2024**, *159*, 127235. [CrossRef]

https://doi.org/10.3390/plants15040544

9. Stebich, M.; Mingram, J.; Han, J.; Liu, J. Late Pleistocene spread of (cool-) temperate forests in Northeast China and climate changes synchronous with the North Atlantic region. *Glob. Planet. Change* **2009**, *65*, 56–70. [CrossRef]
10. Hedges, L.V.; Gurevitch, J.; Curtis, P.S. The meta-analysis of response ratios in experimental ecology. *Ecology* **1999**, *80*, 1150–1156. [CrossRef]
11. Ni, Y.X.; Liu, X.H.; He, L.; Wen, Y.; You, G.-Y. Mobile application-based interventions for people with heart failure: A systematic review and meta-analysis. *J. Nurs. Manag.* **2024**, *2024*, 6859795. [CrossRef] [PubMed]
12. Luo, K.; Yuan, X.; Zhang, K.; Xue, Y.; Fu, Z.; Lin, P.; Li, Y.; Li, Y.; Pu, T.; Qi, X.; et al. Diethyl aminoethyl hexanoate (DA-6) and planting density optimize soybean growth and yield formation in maize–soybean strip intercropping. *Crop. J.* **2025**, *13*, 1259–1270. [CrossRef]
13. Yan, Y.; Duan, F.; Li, X.; Zhao, R.; Hou, P.; Zhao, M.; Li, S.; Wang, Y.; Dai, T.; Zhou, W. Photosynthetic capacity and assimilate transport of the lower canopy influence maize yield under high planting density. *Plant Physiol.* **2024**, *195*, 2652–2667. [CrossRef]
14. Pérez-Ramos, I.M.; Matías, L.; Gómez-Aparicio, L.; Godoy, Ó. Functional traits and phenotypic plasticity modulate species coexistence across contrasting climatic conditions. *Nat. Commun.* **2019**, *10*, 2555. [CrossRef]
15. Tao, Z.; Shen, C.; Qin, W.; Nie, B.; Chen, P.; Wan, J.; Zhang, K.; Huang, W.; Siemann, E. Fluctuations in resource availability shape the competitive balance among non-native plant species. *Ecol. Appl.* **2024**, *34*, e2795. [CrossRef] [PubMed]
16. Zhou, M.; Wang, Y.; Gu, J.; Ma, N.; Hou, W.; Sun, J.; Fan, X.; Yin, G. Effects of increasing maize planting density on yield, water productivity and irrigation water productivity in China: A comprehensive meta-analysis incorporating soil and climatic factors. *Agric. Water Manag.* **2025**, *317*, 109671. [CrossRef]
17. Guo, R.; Liu, L.; Li, J.; Qu, H.; Guo, W.; Zhang, L.; Yang, D.; Wang, R.; Guo, C. Metabolo-transcriptomics analyses reveal alfalfa adaptation to combined saline-alkali and low-temperature stress in the field. *Plant Biotechnol. J.* **2025**, *early view*. [CrossRef]
18. Luo, N.; Meng, Q.; Feng, P.; Qu, Z.; Yu, Y.; Liu, D.L.; Müller, C.; Wang, P. China can be self-sufficient in maize production by 2030 with optimal crop management. *Nat. Commun.* **2023**, *14*, 2637. [CrossRef] [PubMed]
19. Song, H.; Li, B.; Li, Z.J.; Zhang, G.; Lu, X.; Qi, P. Modelling water and land resources synergy and trade-off in a major grain-producing area, China. *Agric. Water Manag.* **2025**, *320*, 109880. [CrossRef]
20. Mudare, S.; Jing, J.; Makowski, D.; He, X.; Liang, Z.; Sims, Z.; Wanger, T.C.; Tilman, D.; Zhang, F.; Cong, W.-F. Crop rotations synergize yield, nutrition, and revenue: A meta-analysis. *Nat. Commun.* **2025**, *16*, 9552. [CrossRef]
21. Wang, X.; Miao, Y.; Dong, L.; Mi, G.; Batchelor, W.D.; Kusnierek, K. Precision nitrogen management for maize based on crop modeling, remote sensing and machine learning. *Comput. Electron. Agric.* **2025**, *239*, 111124. [CrossRef]
22. Ma, R.; Cao, N.; Li, Y.; Hou, Y.; Wang, Y.; Zhang, Q.; Wang, T.; Cui, J.; Li, B.; Shi, W.; et al. Rational reduction of planting density and enhancement of NUE were effective methods to mitigate maize yield loss due to excessive rainfall. *Eur. J. Agron.* **2024**, *160*, 127326. [CrossRef]
23. Wu, F.; Tang, Q.; Cui, J.; Tian, L.; Guo, R.; Wang, L.; Zheng, Z.; Zhang, N.; Zhang, Y.; Lin, T. Deficit irrigation combined with a high planting density optimizes root and soil water–nitrogen distribution to enhance cotton productivity in arid regions. *Field Crops Res.* **2024**, *317*, 109524. [CrossRef]
24. Mueller, N.D.; Gerber, J.S.; Johnston, M.; Ray, D.K.; Ramankutty, N.; Foley, J.A. Closing yield gaps through nutrient and water management. *Nature* **2012**, *490*, 254–257. [CrossRef] [PubMed]
25. Joshi, V.R.; Thorp, K.R.; Coulter, J.A.; Johnson, G.A.; Porter, P.M.; Strock, J.S.; Garcia y Garcia, A. Improving Site-Specific Maize Yield Estimation by Integrating Satellite Multispectral Data into a Crop Model. *Agronomy* **2019**, *9*, 719. [CrossRef]
26. Lajeunesse, M.J. Facilitating systematic reviews, data extraction and meta-analysis with the metagear package for R. *Methods Ecol. Evol.* **2016**, *7*, 323–330. [CrossRef]
27. Lajeunesse, M.J. On the meta-analysis of response ratios for studies with correlated and multi-group designs. *Ecology* **2011**, *92*, 2049–2055. [CrossRef]
28. Viechtbauer, W. Conducting meta-analyses in R with the metafor package. *J. Stat. Softw.* **2010**, *36*, 1–48. [CrossRef]
29. Richardson, M.; Garner, P.; Donegan, S. Interpretation of subgroup analyses in systematic reviews: A tutorial. *Clin. Epidemiol. Glob. Health* **2019**, *7*, 192–198. [CrossRef]
30. Wang, X.; Piantadosi, S.; Le-Rademacher, J.; Mandrekar, S.J. Statistical considerations for subgroup analyses. *J. Thorac. Oncol.* **2021**, *16*, 375–380. [CrossRef]
31. Nakagawa, S.; Santos, E.S.A. Methodological issues and advances in biological meta-analysis. *Evol. Ecol.* **2012**, *26*, 1253–1274. [CrossRef]

 https://doi.org/10.3390/plants15040544

MDPI AG
Grosspeteranlage 5
4052 Basel
Switzerland
Tel.: +41 61 683 77 34

Plants Editorial Office
E-mail: plants@mdpi.com
www.mdpi.com/journal/plants

www.ingramcontent.com/pod-product-compliance
Lightning Source LLC
Chambersburg PA
CBHW040717270926
51850CB00023B/2676
9783725873609